Lecture Notes in Mathematics

Edited by A. Dold and B. Eckmann

1325

A. Truman I. M. Davies (Eds.)

Stochastic Mechanics and Stochastic Processes

Proceedings of a Conference
held in Swansea, U.K., Aug. 4–8, 1986

Springer-Verlag
Berlin Heidelberg New York London Paris Tokyo

Editors

Aubrey Truman
Ian M. Davies
Department of Mathematics,
University College of Swansea
Singleton Park, Swansea SA2 8PP
Wales, U.K.

Mathematics Subject Classification (1980): 60GXX, 60HXX

ISBN 3-540-50015-4 Springer-Verlag Berlin Heidelberg New York
ISBN 0-387-50015-4 Springer-Verlag New York Berlin Heidelberg

Printing and binding: Druckhaus Beltz, Hemsbach/Bergstr.
2146/3140-543210

Preface

This volume primarily contains papers presented at the meeting entitled 'Stochastic Mechanics and Stochastic Processes' held in Swansea from 4 August to 8 August 1986. Also included in the volume are some related papers, not presented at the meeting, but in the same subject area. The topics covered herein are quite varied but all related to the central themes of the meeting, including large deviations and statistical mechanics, Nelson's stochastic mechanics and quantum diffusions, simulations of Brownian Motions and stochastic flows. The meeting was most worthwhile both in the quality of the talks given and the level of discussion generated on the subject of stochastic processes and stochastic mechanics.

Most of the papers herein are reasonably self-contained and should be readily accessible to researchers in this field. For beginning students in Nelson's stochastic mechanics we recommend that they start by reading the first paper in this volume by Batchelor and Truman, which deals with stochastic mechanics for excited states of a system with finitely many degrees of freedom. Some of the corresponding stochastic flows are discussed in the paper by Chappell and Elworthy, where new results are given for their Lyapunov exponents. There are related papers by Yasue, Durran and Williams and Steele. Zambrini's paper includes new results for Bernstein processes and stochastic mechanics. An interesting new treatment of stochastic mechanics for systems with infinitely many degrees of freedom (quantum fields) is given in the paper by Carlen. Supersymmetry is discussed in the paper by Haba.

There is an excellent expository account of large deviations in statistical mechanics by Lewis, and two papers by Lewis and co-workers in statistical mechanics itself. There have been exciting new developments in this area recently. An introduction to the algebraic theory of quantum diffusions is given in the paper by Hudson. New analytical results for stochastic processes are given in the papers by Kifer, McGill, McGregor and Watling.

It is a pleasure to thank Brian Davies, of King's College London, and Nick Bingham, of Royal Holloway, for their assistance in helping to organise the meeting and in producing this volume. We are grateful to the SERC for financial support through research grant GR/D/88847. We should like to thank Mrs E. Evans, Mrs M. Prowse and Mrs E. Williams for their patience and application in typing some of the papers. Last but far from least we would like to thank the referee for his important contribution.

A. Truman and I.M. Davies
Swansea

CONTENTS

ON FIRST HITTING TIMES IN STOCHASTIC MECHANICS

by

Andrew Batchelor and Aubrey Truman
Department of Mathematics and Computer Science
University College of Swansea
Singleton Park, Swansea SA2 8PP

1. A Resumé of Stochastic Mechanics

The Schrödinger equation for a particle Q of mass m subject to a force $-\underset{\sim}{\nabla}V$ in $\mathbb{R}^d$ is equivalent to

$$i\hbar\psi^* \frac{\partial\psi}{\partial t} = -\frac{\hbar^2}{2m}\psi^*\Delta_x\psi + \psi^* V\psi, \tag{1}$$

where $*$ denotes the complex conjugate and $\hbar$ is Planck's constant divided by 2π. Here $\psi = \psi(\underset{\sim}{x},t)$ is the quantum mechanical wave-function and the time $t \in \mathbb{R}^+$, the positive reals, and $\underset{\sim}{x} \in \mathbb{R}^d$, d-dimensional Euclidean configuration space.

Since V is real-valued, equating imaginary parts of the above equation gives the continuity equation

$$\frac{\partial\rho}{\partial t} + \operatorname{div} \underset{\sim}{j} = 0, \tag{2}$$

where $\rho = \psi^*\psi$ is the quantum mechanical particle density and $\underset{\sim}{j} = \frac{\hbar}{2mi}(\psi^*\underset{\sim}{\nabla}\psi - \psi\underset{\sim}{\nabla}\psi^*)$ is the probability current. The last equation merely expresses the conservation of particle number in that, for the state ψ, the probability that Q is in A at time t is given by

$$\mathbb{P}\{Q \in A \text{ at time } t\} = \int_A \rho(\underset{\sim}{x},t)dx, \tag{3}$$

for each Borel set $A \subset \mathbb{R}^d$.

Following Nelson [6],[7], introduce the real-valued functions R and S defined by $\psi = e^{R+iS}$, so that $\rho = e^{2R}$ and $\underset{\sim}{j} = (\frac{\hbar}{m}\underset{\sim}{\nabla} S)e^{2R}$. We can then deduce from the continuity equation that

$$\frac{\partial\rho}{\partial t} = \operatorname{div}(-\frac{\hbar}{m}\underset{\sim}{\nabla} S e^{2R}) = \operatorname{div}(\frac{\hbar}{2m}\underset{\sim}{\nabla} e^{2R} - \frac{\hbar}{m}\underset{\sim}{\nabla}(R+S)e^{2R}) \tag{4}$$

i.e.

$$\frac{\partial\rho}{\partial t} = \operatorname{div}(\nu\underset{\sim}{\nabla}\rho - \underset{\sim}{b}\rho), \tag{5}$$

for $\nu = \frac{\hbar}{2m}$ and $\underset{\sim}{b} = \frac{\hbar}{m}\underset{\sim}{\nabla}(R+S)$.

The last equation is just the forward Kolmogorov equation for the density ρ for the diffusion $\underset{\sim}{X}$ satisfying

$$d\underset{\sim}{X} = \underset{\sim}{b}(\underset{\sim}{X}(t),t)dt + \left(\frac{\hbar}{m}\right)^{\frac{1}{2}} d\underset{\sim}{B}(t), \tag{6}$$

where the forward drift $\underset{\sim}{b} = \frac{\hbar}{m}\underset{\sim}{\nabla}(R+S)$, $\underset{\sim}{B} = (B_1, B_2, \ldots, B_d)$ in cartesians, with $\mathbb{E}\{B_i(t)B_j(s)\} = \delta_{ij}\min(s,t)$, for $i,j = 1,2,\ldots,d$, $\underset{\sim}{B}$ being a $BM(\mathbb{R}^d)$ process. Moreover, equating the real parts of Eq. (1) gives for $\psi = e^{R+iS}$

$$\frac{\partial S}{\partial t} = \frac{\hbar}{2m}(|\underset{\sim}{\nabla}R|^2 - |\underset{\sim}{\nabla}S|^2 + \Delta R) - \frac{V}{\hbar}. \tag{7}$$

Nelson's remarkable discovery was that the last equation embodies a dynamical principle for the diffusion $\underset{\sim}{X}$.

To see this Nelson defines the mean forward and backward time derivatives $D_\pm$ by

$$D_\pm f(\underset{\sim}{X}(t),t) = \lim_{h\downarrow 0} \mathbb{E}\left\{\frac{f(\underset{\sim}{X}(t\pm h),t\pm h) - f(\underset{\sim}{X}(t),t)}{\pm h}\middle|\underset{\sim}{X}(t)\right\}. \tag{8}$$

Then it follows from Eq. (6) that

$$D_+\underset{\sim}{X}(t) = \underset{\sim}{b}(\underset{\sim}{X}(t),t) = \frac{\hbar}{m}\underset{\sim}{\nabla}(S+R)(\underset{\sim}{X}(t),t) \tag{9}$$

and from Itô's formula that for sufficiently regular f

$$D_+ f(\underset{\sim}{X}(t),t) = \left(\frac{\partial f}{\partial t} + \underset{\sim}{b}.\underset{\sim}{\nabla}f + \frac{\hbar}{2m}\Delta f\right)(\underset{\sim}{X}(t),t). \tag{10}$$

Nelson went on to show that for sufficiently regular functions g and h

$$\frac{d}{dt}\mathbb{E}(g(\underset{\sim}{X}(t))h(\underset{\sim}{X}(t))) = \mathbb{E}(g(\underset{\sim}{X}(t))D_-h(\underset{\sim}{X}(t))) + \mathbb{E}(D_+g(\underset{\sim}{X}(t))h(\underset{\sim}{X}(t))) \tag{11}$$

(see e.g. p. 98 Ref (6)).

We can now establish Nelson's result. Firstly, from above, for any $f \in C_0^\infty(\mathbb{R}^d)$, for $i = 1,2,\ldots,d$,

$$\frac{d}{dt}\mathbb{E}\{f(\underset{\sim}{X}(t))X_i(t)\} = \mathbb{E}\{f(\underset{\sim}{X}(t))D_-X_i(t)\} + \mathbb{E}\{X_i(t)(\frac{\hbar}{m}(\underset{\sim}{\nabla}R + \underset{\sim}{\nabla}S).\underset{\sim}{\nabla}f + \frac{\hbar}{2m}\Delta f)(\underset{\sim}{X}(t),t)\}, \tag{12}$$

where $\underset{\sim}{X} = (X_1, X_2, \ldots, X_d)$ in cartesians. Since $\mathbb{E}(g(\underset{\sim}{X}(t),t)) = \int_{\mathbb{R}^d} e^{2R(\underset{\sim}{x},t)} g(\underset{\sim}{x},t)dx$, integrating by parts, using the identity, for $i = 1,2,\ldots,d$,

$$-e^{-2R(\underset{\sim}{x},t)}\operatorname{div}\{x_i\underset{\sim}{\nabla}R(\underset{\sim}{x},t)e^{2R(\underset{\sim}{x},t)}\} + 2^{-1}e^{-2R(\underset{\sim}{x},t)}\Delta(x_i e^{2R(\underset{\sim}{x},t)}) = \nabla_i R(\underset{\sim}{x},t), \tag{13}$$

for $\underset{\sim}{\nabla}R = (\nabla_1 R, \nabla_2 R, \ldots, \nabla_d R)$ in cartesians, we obtain for each $f \in C_0^\infty(\mathbb{R}^d)$

$$\frac{d}{dt}\mathbb{E}\{f(\underset{\sim}{X}(t))X_i(t)\} = \mathbb{E}\{f(\underset{\sim}{X}(t)D_-X_i(t)\} + \frac{\hbar}{m}\mathbb{E}\{X_i(t)(\underset{\sim}{\nabla}S.\underset{\sim}{\nabla}f)(\underset{\sim}{X}(t),t)\} + \frac{\hbar}{m}\mathbb{E}\{(\nabla_i R)f(\underset{\sim}{X}(t),t)\}. \tag{14}$$

But Ehrenfest's theorem for the quantum mechanical state $\psi = e^{R+iS}$ gives for sufficiently regular g

$$\frac{d}{dt}\mathbb{E}(g(\underset{\sim}{X})) = \frac{\hbar}{m}\mathbb{E}((\underset{\sim}{\nabla}g).(\underset{\sim}{\nabla}S)(\underset{\sim}{X},t)). \tag{15}$$

Hence, setting $g(\underset{\sim}{X}) = f(\underset{\sim}{X})X_i$, we obtain

$$\frac{d}{dt}\mathbb{E}(f(\underset{\sim}{X})X_i) = \frac{\hbar}{m}\mathbb{E}(\nabla_i S(\underset{\sim}{X},t)f(\underset{\sim}{X})) + \frac{\hbar}{m}\mathbb{E}(X_i\underset{\sim}{\nabla}S(\underset{\sim}{X},t).\underset{\sim}{\nabla}f(X)), \tag{16}$$

for $i = 1,2,\ldots,d$.

Comparing this with Eq. (14) above, we see that necessarily

$$D_-X_i(t) = \frac{\hbar}{m}(\nabla_i S - \nabla_i R)(\underset{\sim}{X}(t),t) \tag{17}$$

for $i = 1,2,\ldots,d$. Hence, the backward drift $D_-\underset{\sim}{X}(t)$ is given by

$$D_-\underset{\sim}{X}(t) = \frac{\hbar}{m}(\underset{\sim}{\nabla}S - \underset{\sim}{\nabla}R)(\underset{\sim}{X}(t),t) \tag{18}$$

and from Itô's formula

$$D_-f(\underset{\sim}{X}(t),t) = (\frac{\partial f}{\partial t} + \frac{\hbar}{m}(\underset{\sim}{\nabla}S - \underset{\sim}{\nabla}R).\nabla f - \frac{\hbar}{2m}\Delta f)(\underset{\sim}{X}(t),t). \tag{19}$$

Nelson's amazing discovery now follows from Eqs. (9), (10), (18) and (19). After a tedious calculation we obtain

$$\frac{m}{2}(D_-D_+ + D_+D_-)\underset{\sim}{X}(t) = \{\hbar\underset{\sim}{\nabla}\frac{\partial S}{\partial t} - \frac{\hbar^2}{2m}\underset{\sim}{\nabla}(|\underset{\sim}{\nabla}R|^2 - |\underset{\sim}{\nabla}S|^2 + \Delta R)\}(\underset{\sim}{X}(t),t), \tag{20}$$

or from Eq. (7)

$$\frac{m}{2}(D_-D_+ + D_+D_-)\underset{\sim}{X}(t) = -\underset{\sim}{\nabla}V(\underset{\sim}{X}(t)). \tag{21}$$

This is the Nelson-Newton law i.e. a stochastic version of Newton's second law of motion

$$\text{Force} = \text{Mass} \times \text{Acceleration}. \tag{22}$$

Therefore, we have seen that the net content of the Schrödinger equation for the state $\psi = e^{R+iS}$ is just the Kolmogorov equation for the diffusion $\underset{\sim}{X}$ with drift $\frac{\hbar}{m}(\underset{\sim}{\nabla}R + \underset{\sim}{\nabla}S)$ and the dynamical principle for $\underset{\sim}{X}$ contained in Eq. (21). This suggests that the sample paths of the diffusion $\underset{\sim}{X}$ have some physical significance. We investigate this for the stationary states of the Coulomb problem below presenting some new results for first hitting times. We do not give all the details of the proofs here and we refer the reader to the original references for details. (See

Refs. (2) and (3)).

2. First Hitting Times for Ground States of Spherically Symmetric Potentials

We consider Nelson diffusions corresponding to ground state solutions of the Schrödinger equation for a quantum mechanical Hamiltonian $H = (-\frac{\hbar^2}{2m}\Delta + V)$, a self-adjoint linear operator on some appropriate domain in $L^2(\mathbb{R}^d)$. We specialize to the case $d = 3$ and further we assume that the potential V is spherically symmetric, $V = V(|\underset{\sim}{x}|)$, for $\underset{\sim}{x} \in \mathbb{R}^3$, $|\cdot|$ being the Euclidean norm.

Let $\Psi_E(\underset{\sim}{x},t) = \Psi_t^E(\underset{\sim}{x}) \in L^2(\mathbb{R}^3)$ be such a ground state with

$$\Psi_E(\underset{\sim}{x},t) = |\underset{\sim}{x}|^{-1} f_E(|\underset{\sim}{x}|)e^{-iEt/\hbar}, \tag{23}$$

E being inf spec (H). Then, as is well-known, f_E satisfies $(H_r - E)f_E = 0$, i.e. setting $x = |\underset{\sim}{x}|$,

$$\{-\frac{\hbar^2}{2m}\frac{d^2}{dx^2} + V(x) - E\}f_E(x) = 0, \tag{24}$$

where $H_r = -\frac{\hbar^2}{2m}\frac{d^2}{dx^2} + V(x)$ is the radial Hamiltonian. We assume that V is piecewise continuous with finite discontinuities on $(0,\infty)$ so that f_E is C^2 and, of course, $f_E > 0$ on $(0,\infty)$ and $f_E(0) = 0$.

A straight-forward application of Itô's formula yields for the Nelson diffusion $\underset{\sim}{X}$ corresponding to the state Ψ_E

$$d|\underset{\sim}{X}| = \frac{\hbar}{m}\frac{d}{d|\underset{\sim}{X}|}\ln f_E(|X|)\,dt + (\frac{\hbar}{m})^{\frac{1}{2}}d\beta(t), \tag{25}$$

β being a $BM(\mathbb{R})$ process. This is a one-dimensional time-homogeneous diffusion with generator

$$L_x = \frac{\hbar}{2m}\frac{d^2}{dx^2} + \frac{\hbar}{m}\frac{d}{dx}\ln f_E(x)\frac{d}{dx}\,. \tag{26}$$

By virtue of Eq. (24) the radial diffusion $|\underset{\sim}{X}|$ satisfies the Nelson-Newton law

$$\frac{m}{2}(D_+D_- + D_-D_+)|\underset{\sim}{X}(t)| = -\frac{d}{d|X|}V(|X|(t)). \tag{27}$$

Moreover, since for any C^2 function f,

$$f_E^{-1}(H_r - E)(f_E f) = f_E^{-1}\{-\frac{\hbar^2}{2m}(f_E f'' + 2f_E' f' + f_E'' f) + (V-E)f_E f\} = -\hbar Lf, \tag{28}$$

f_E being positive, we obtain formally at least

$$p_t(x,y) = f_E^{-1}(x)\exp\{-\frac{t}{\hbar}(H_r - E)\}(x,y)f_E(y), \tag{29}$$

p_t being the transition density for $|\underset{\sim}{X}|$, $\exp\{-\frac{t}{\hbar}(H_r - E)\}(x,y)$ being the appropriate heat kernel. For convenience now set $\hbar = m = 1$.

We now set about finding the distribution of $\tau_x(a)$:

$$\tau_x(a) = \inf\{s > 0 : |\underset{\sim}{X}(s)| = a,\ |\underset{\sim}{X}(0)| = x\}, \tag{30}$$

the first hitting time of the level a for the process starting at x. The key result here is

$$\mathbb{E}\{\exp(-\lambda\tau_x(a))\} = f_E^{-1}(x)\,\frac{(H_r + \lambda - E)^{-1}(x,a)}{(H_r + \lambda - E)^{-1}(a,a)}\,f_E(a), \qquad (\lambda > 0), \tag{31}$$

where $(H_r + \lambda - E)^{-1}(x,a)$ is the resolvent kernel. The last identity follows because $|\underset{\sim}{X}|$ is a one-dimensional time-homogeneous diffusion. For such a diffusion to have gone from x to y in time t the first hitting time of any intermediate point a must be less than t. Since the process starts afresh from a, for each fixed x,y and any intermediate a,

$$p_t(x,y) = \int_0^t g(x,a;u)p_{t-u}(a,y)du, \tag{32}$$

where $g(x,a;u)du = \mathbb{P}\{\tau_x(a) \in (u,u+du)\}$. The desired identity follows by taking Laplace transforms and letting $y \to a$. This leads to:

<u>PROPOSITION 1</u>

Let $E(<0)$ be inf spec (H), where $H = -2^{-1}\Delta_x + V(|\underset{\sim}{x}|)$. For each $\lambda > 0$, let the equation

$$-2^{-1}\frac{d^2y}{dx^2} + (V(x) + \lambda - E)y = 0 \tag{33}$$

have two linearly independent C^2 solutions $f_{E-\lambda}$ and $g_{E-\lambda}$ defined by $f_{E-\lambda}(0) = 0$, $f'_{E-\lambda}(0) = 1$, $g_{E-\lambda}(0) = 1$, $g'_{E-\lambda}(0) = 0$. (This will be the case e.g. if $x^2V(x)$ is analytic in a neighbourhood of the origin.) Then, if V is piecewise continuous, with $V(x) \to 0$ as $x \to \infty$, for $\tau_x(a)$ the first hitting time for the ground state radial process, for the Hamiltonian H, for $x < a$,

$$\mathbb{E}\{\exp(-\lambda\tau_x(a))\} = f_{E-\lambda}(x)f_E(a)/f_{E-\lambda}(a)f_E(x), \tag{34}$$

whilst for $x > a$

$$\mathbb{E}\{\exp(-\lambda\tau_x(a))\} = h_{E-\lambda}(x)f_E(a)/h_{E-\lambda}(a)f_E(x), \tag{35}$$

h being the unique solution of Eq. (33), which is exponentially decreasing at infinity.

The above result yields:

COROLLARY 1

Let $x > a$ and further let V have compact support with $\operatorname{supp} V \subset [0,a)$. Then for the radial ground state process with energy $E(<0)$

$$\mathbb{E}\{\exp(-\lambda\tau_x(a))\} = \exp(-\surd(-2(E-\lambda))(x-a))\exp(+\surd(-2E)(x-a)). \tag{36}$$

Thus,

$$\mathbb{P}(\tau_x(a) \in ds) = (2\pi)^{-\frac{1}{2}}|a-x|s^{-3/2}\exp\{Es - \frac{(x-a)^2}{2s} + \surd(-2E)(x-a)\}ds. \tag{37}$$

In this case, therefore, for $x > a$, with $\operatorname{supp} V \subset [0,a)$, the distribution of $\tau_x(a)$ is the same as that for Brownian motion with a constant drift $-\surd(-2E)$ (see e.g. Ref. (10)). The last example is atypical in that one has an explicit formula for the distribution of $\tau_x(a)$. More typical is the next example.

Example 1 (Spherical Square Well Ground State)

Let $V(\underset{\sim}{x}) = -V_o$, for $|\underset{\sim}{x}| < r_o$, $V(\underset{\sim}{x}) = 0$, otherwise. The ground state energy E and ground state wave-function f_E are defined by

$$f_E(|\underset{\sim}{x}|) = \begin{cases} \sinh(\surd(-2(E+V_o))|\underset{\sim}{x}|), & |\underset{\sim}{x}| < r_o, \\ B\exp(-\surd(-2E))|\underset{\sim}{x}|), & |\underset{\sim}{x}| > r_o, \end{cases}$$

where E and B satisfy

$$\sinh(\surd(-2(E+V_o))r_o) = B\exp(-\surd(-2E)r_o),$$

$$\frac{d}{dr_o}\sinh(\surd(-2(E+V_o))r_o) = \frac{d}{dr_o}B\exp(-\surd(-2E)r_o).$$

Let $x < a < r_o$. Then

$$\mathbb{E}\{\exp(-\lambda\tau_x(a))\} = \frac{\sinh(\surd(-2(E+V_o-\lambda))x)\sinh(\surd(-2(E+V_o))a)}{\sinh(\surd(-2(E+V_o-\lambda))a)\sinh(\surd(-2(E+V_o))x)}. \tag{38}$$

Similar results can be obtained for other values of x and a. A more interesting physical example is:

Example 2 (Coulomb Problem Ground State)

Let $V(\underset{\sim}{x}) = -Ze^2/|\underset{\sim}{x}|$. The ground state energy $E = -1/2a_o^2$, a_o being the Bohr radius $1/Ze^2$. The corresponding ground state wave-function is $f_E(|\underset{\sim}{x}|) = |\underset{\sim}{x}|\exp(-|\underset{\sim}{x}|/a_o)$. In this case we obtain

$$\mathbb{E}\{\exp(-\lambda\tau_r(a))\} = \left(\frac{a}{r}\right) \exp\left(\frac{r-a}{a_o}\right) \frac{W_{\frac{1}{\kappa},\frac{1}{2}}\left(\frac{2\kappa r}{a_o}\right)}{W_{\frac{1}{\kappa},\frac{1}{2}}\left(\frac{2\kappa a}{a_o}\right)} \quad (r \geq a), \tag{39}$$

and

$$\mathbb{E}\{\exp(-\lambda\tau_r(a))\} = \left(\frac{a}{r}\right) \exp\left(\frac{r-a}{a_o}\right) \frac{M_{\frac{1}{\kappa},\frac{1}{2}}\left(\frac{2\kappa r}{a_o}\right)}{M_{\frac{1}{\kappa},\frac{1}{2}}\left(\frac{2\kappa a}{a_o}\right)} \quad (r \leq a), \tag{40}$$

a_o being the Bohr radius and $\kappa = \sqrt{(1+2a_o^2\lambda)}$, W and M being Whittaker functions. (See Ref. (3)).

Needless to say we cannot find the inverse Laplace transforms for either of above examples. We choose a different approach below.

3. Expected Values of First Hitting Times

We begin with a minor extension of an analytical result of Mandl (see Ref. (5)). We use the same notation as above.

PROPOSITION 2

Let $f_E(>0)$ be such that $b(x) = f_E'(x)/f_E(x)$ is Lipschitz continuous on $(0,\infty)$, 0 and ∞ being entrance and natural boundaries for the radial gound-state diffusion $|\underset{\sim}{X}|$ with generator $L_x = 2^{-1}\frac{d^2}{dx^2} + b(x)\frac{d}{dx}$. Then for the radial ground state process $v(x) = \mathbb{E}(\tau_x(a))$ satisfies

$$L_x v(x) = 2^{-1}v''(x) + b(x)v'(x) = -1, \tag{41}$$

together with the boundary conditions $v(a) = 0$ and $\lim_{x\downarrow 0} f_E^2(x)v'(x) = 0$ for $x < a$ and $\lim_{x\uparrow\infty} f_E^2(x)v'(x) = 0$ for $x > a$.

The proof of this relies heavily on the methods of Mandl (see Refs. (5) and (3)). Here we content ourselves with a formal explanation of what is going on. First we differentiate with respect to λ the equation

$$(H_r + \lambda - E)f_{E-\lambda} = 0,$$

yielding

$$\frac{df}{d\lambda}E - \lambda = -(H_r + \lambda - E)^{-1} f_{E-\lambda}. \tag{42}$$

If we assume $v(x) = \mathbb{E}(\tau_x(a)) = \lim_{\lambda \downarrow 0} \frac{d}{d\lambda}\mathbb{E}\{\exp(-\lambda\tau_x(a))\}$ the last identity leads to the desired conditions on v.

To see this refer back to the results of the last section for $x < a$, using $f_{E-\lambda} \to f_E$ as $\lambda \to 0$,

$$-\lim_{\lambda \downarrow 0} \frac{d}{d\lambda}\mathbb{E}\{\exp(-\lambda\tau_x(a))\} = \frac{(H_r - E)^{-1} f_E(x)}{f_E(x)} - \frac{(H_r - E)^{-1} f_E(a)}{f_E(a)}, \tag{43}$$

where $(H_r - E)^{-1}$ is an integral operator whose boundary conditions are to be found. Recalling that $L = -f_E^{-1}(H_r - E)f_E$, we obtain

$$L_x v(x) = -1, \quad v(a) = 0, \tag{44}$$

for $v(x) = \mathbb{E}(\tau_x(a))$. The remaining boundary condition comes about because

$$v(x) = \mathbb{E}(\tau_x(a)) = -\frac{\frac{df}{d\lambda}E - \lambda\Big|_{\lambda=0}(x)}{f_E(x)} + \frac{\frac{df}{d\lambda}E - \lambda\Big|_{\lambda=0}(a)}{f_E(a)}. \tag{45}$$

Differentiating, using the facts that $f_E(x) \to 0$ and $f_E'(x) \to 1$ as $x \to 0$, gives formally at least, for $x < a$

$$\lim_{x \downarrow 0} f_E^2(x) v'(x) = \lim_{x \downarrow 0} \frac{df}{d\lambda}E - \lambda\Big|_{\lambda=0}(x) = 0, \tag{46}$$

for a sufficiently smooth $f_{E-\lambda}(x)$ in (λ, x) in a neighbourhood of $(0,0)$, since $f_{E-\lambda}(0+) = 0$. (For further details see Ref. (3) and Batchelor's Ph.D. thesis.)

The last proposition gives for the Coulomb problem:

PROPOSITION 3

For the radial ground state diffusion for the Coulomb potential $V = -Ze^2/|\underset{\sim}{x}|$

$$\mathbb{E}(\tau_r(a)) = a_o^2 \int_{a/a_o}^{r/a_o} (1 + \frac{1}{x} + \frac{1}{2x^2})\, dx \quad (r \geq a), \tag{47}$$

and

$$\mathbb{E}(\tau_r(a)) = a_o^2 \int_{r/a_o}^{a/a_o} \{\frac{1}{2x^2}(e^{2x} - 1 - 2x) - 1\}\, dx \quad (r \leq a), \tag{48}$$

$a_o = 1/Ze^2$ being the Bohr radius. Further

$\mathbb{E}(\tau_r(a) | r \in (a,\infty)$ is distributed according to quantum mechanical ground state distribution)

$$= a_o^2 \int_{a/a_o}^{\infty} (x + 1 + \frac{1}{2x})^2 e^{-2x} dx \qquad (49)$$

and

$\mathbb{E}(\tau_r(a) | r \in (0,a)$ is distributed according to quantum mechanical ground state distribution)

$$= a_o^2 \int_0^{a/a_o} \{\frac{\sinh x}{x} - (1+x)e^{-x}\}^2 dx. \qquad (50)$$

The question arises as to whether one can check any of the above results experimentally. Here we reiterate an idea in Ref. (9).

A Possible Experimental Test

If you replace the negatively charged electron in the hydrogen atom by a negatively charged pion π^-, the π^- feels only the Coulomb attraction due to the positively charged proton p^+ when at a distance exceeding $\hbar/m_\pi c$ (the pion Compton wavelength) from the nuclear proton. The reason is that the strong force governing the decay $p^+ + \pi^- \to n + \gamma$'s is of extremely short range $\sim \hbar/m_\pi c$. Therefore the $p^+\pi^-$ system cannot decay until the π^- first hits a sphere of radius $\hbar/m_\pi c$ centred at the nuclear proton.

Assume therefore that the π^- is captured into the ground state. Then, if stochastic mechanics gives the correct first hitting time, we obtain

$$\text{Expected Decay Time for Ground State of } p^+\pi^- > \frac{2\hbar^3}{m_\pi e^4} \int_{e^2/c\hbar}^{\infty} (x + 1 + \frac{1}{2x})^2 e^{-2x} dx \sim 10^{-18} \text{ secs.} \qquad (51)$$

Experiment gives:

$$\text{Expected Decay Time for } p^+\pi^- < 10^{-12} \text{ secs,}$$

which is consistent. This begs the question as to whether or not one can determine the expected first hitting times for excited states.

Denote by $a_{1,1}$ the first zero of the radial wave function for the first excited state for the Coulomb problem, i.e. the first zero of L_1^1, the Laguerre polynomial.

We let $\bar{\mathbb{E}}^1_+(\tau(a))$ denote $\mathbb{E}(\tau_r(a) | r \in (a, a_{1,1})$ is distributed according to the probability distribution of the first excited state) and $\bar{\mathbb{E}}^1_-(\tau(a))$ the corresponding expression for a diffusion initially distributed according to probability distribution of first excited state on $(0,a)$. Then the methods above yield:

PROPOSITION 4

For the first excited state for the Coulomb potential $-Ze^2/|\underset{\sim}{x}|$

$$\bar{\mathbb{E}}^1_+(\tau(a)) = \frac{8a_o^2}{(1-7e^{-2})} \int_{a/2a_o}^{1} \frac{1}{(1-\rho)^2} \{(\rho^3+\rho+1+\frac{1}{2\rho})e^{-\rho} - \frac{7e^{-(2-\rho)}}{2\rho}\}^2 d\rho \tag{52}$$

and

$$\bar{\mathbb{E}}^1_-(\tau(a)) = \frac{8a_o^2}{(1-7e^{-2})} \int_0^{a/2a_o} \frac{1}{(1-\rho)^2} \{\frac{\sinh\rho}{\rho} - (\rho^3+\rho+1)e^{-\rho}\}^2 d\rho . \tag{53}$$

Further details of the above calculation are given in Batchelor's Ph.D. thesis.

It is clearly going to be very difficult to compute analogues of the above for the nth excited state. We therefore resort to asymptotic methods. (See Refs. (5) and (8)).

4. Asymptotic Results

The nth excited state, with zero angular momentum, for the Coulomb problem has radial wave-function

$$f_E(x) = x \exp\left(\frac{-x}{(n+1)a_o}\right) L_n^1\left(\frac{2x}{(n+1)a_o}\right) \qquad n = 0,1,2,\ldots \tag{54}$$

$a_o = 1/Ze^2$ being the Bohr radius, L_n^1 a Laguerre polynomial, the energy level $E = E_n = -\frac{Z^2e^4}{2(n+1)^2}$. The zeros of the Laguerre polynomial, labelled by $a_{i,n}$ in increasing order, are unattainable points so internodal regions $(a_{i,n}, a_{i+1,n})$ are non-communicating (see e.g. Ref. (1)). We work with wave-function f_E on $(0, a_{1,n})$ to find first hitting time of $a \sim 0$ for each energy level E_n for zero angular momentum.

We follow the methods of Mandl and Newell (see Refs. (5) and (8)) and seek a $\gamma(a) \uparrow \infty$ as $a \downarrow 0$, with

$$\lim_{a\downarrow 0} \mathbb{E}\{\exp(-\frac{\lambda\tau_x(a)}{\gamma(a)})\} = (1+\lambda)^{-1}, \tag{55}$$

for each fixed $\lambda > 0$. That such a γ exists can be seen from the results of the last section.

From Eq. (35), since $\lim_{\lambda\downarrow 0} h_{E-\lambda}(x) = f_E(x)$, for $x > a$, we obtain

$$\lim_{a\downarrow 0} \mathbb{E}\{\exp(-\frac{\lambda\tau_x(a)}{\gamma(a)})\} = \lim_{a\downarrow 0} \frac{f_E(a)}{h_{E-\frac{\lambda}{\gamma(a)}}(a)} \tag{56}$$

and from de l'Hopital's rule

$$\lim_{a\downarrow 0} \mathbb{E}\{\exp(-\frac{\lambda\tau_x(a)}{\gamma(a)})\} = \lim_{a\downarrow 0} \frac{f_E'(a)}{\left[h'_{E-\frac{\lambda}{\gamma(a)}}(a) - \lambda \frac{dh_{E-\mu}}{d\mu}\Big|_{\mu=0}(a)\,\frac{\gamma'(a)}{\gamma^2(a)}\right]}. \tag{57}$$

Hence, we obtain

$$\lim_{a\downarrow 0} \mathbb{E}\{\exp(-\frac{\lambda\tau_x(a)}{\gamma(a)})\} = (1+\lambda)^{-1}, \tag{58}$$

if γ is defined to satisfy

$$\frac{\frac{dh_{E-\mu}}{d\mu}\Big|_{\mu=0}(a)\,\frac{\gamma'(a)}{\gamma^2(a)}}{f_E'(a)} = -1. \tag{59}$$

The following is a minor extension of a result of Mandl and Newell (see Refs. (5) and (8)).

PROPOSITION 5

For the radial Nelson diffusion corresponding to the nth eigenvalue E_n and eigenfunction f_{E_n}, restricted to $(0,a_{1,n})$, $\gamma(a) = 2\int_0^{a_{1,n}} f_{E_n}^2(x)\,dx \int^a f_{E_n}^{-2}(y)\,dy$

and for each fixed x

$$\lim_{a\downarrow 0} \mathbb{E}\{\exp(-\frac{\lambda\tau_x(a)}{\gamma(a)})\} = (1+\lambda)^{-1}.$$

Let us see why above γ is consistent with our Eq. (59). Firstly, if

$\lim_{a\downarrow 0} \mathbb{E}\{\exp(-\frac{\lambda\tau_x(a)}{\gamma(a)})\} = (1+\lambda)^{-1}$, for small λ, it is necessary that, as $a \sim 0$,

$$\gamma(a) \sim \mathbb{E}\{\tau_x(a)\} = \frac{-\frac{dh}{d\lambda}E{-}\lambda\Big|_{\lambda=0}(x)}{f_E(x)} + \frac{\frac{dh}{d\lambda}E{-}\lambda\Big|_{\lambda=0}(a)}{f_E(a)} . \tag{60}$$

Therefore, our equation reduces to

$$\frac{\gamma'(a)}{\gamma^2(a)} \sim -\frac{f_E'(a)}{\frac{dh}{d\lambda}E{-}\lambda\Big|_{\lambda=0}(a)} \sim -\frac{f_E'(a)}{f_E(a)(\gamma(a)+c(x))}, \quad \text{as} \quad a \sim 0, \tag{61}$$

for a function $c(x)$. Evidently this equation is incapable of determining the multiplicative constant $2\int_0^{a_{1,n}} f_E^2(x)\,dx$. However, observe that $\gamma(a) = \int^a f_E^{-2}(x)\,dx$ gives

$$\frac{\gamma'(a)}{\gamma^2(a)} = \frac{f_E^{-2}(a)}{(\int^a f_E^{-2}(x)dx)^2} \sim -\frac{2f^{-3}(a)f_E'(a)}{2\int^a f_E^{-2}(x)\,dx\, f_E^{-2}(a)} = -\frac{f_E'(a)}{f_E(a)\gamma(a)} , \tag{62}$$

as it should, since $f_E(a) \sim 0$. The multiplicative constant is determined by Eq. (60). Formally then for the radial diffusion corresponding to the nth eigenfunction f_{E_n}

$$\mathbb{P}(\tau_x(a) < t) \sim 1 - e^{-t/\gamma_n(a)} , \tag{63}$$

where $\gamma_n(a) = 2\int_0^{a_{1,n}} f_{E_n}^2(x)\,dx \int^a f_{E_n}^{-2}(y)\,dy$. The following proposition was discovered by Andrew Batchelor on the computer.

PROPOSITION 6

$$\lim_{n\uparrow\infty} \lim_{a\downarrow 0} \frac{\gamma^n(a)}{\gamma^0(a)} = \frac{1}{16}\int_0^{j_{1,1}} x^3(J_1(x))^2 dx, \tag{64}$$

$j_{1,1}$ being first zero of first Bessel function J_1.

Further details are given in Refs. (2) and (3). Batchelor's computer tabulation is given at the end of this paper. There is a corresponding proposition for non-zero angular momentum states:

$$f_{n,\ell}(x) = x^{\ell+1} L_n^{2\ell+1}\left(\frac{2x}{a_o(n+\ell+1)}\right) e^{-x/(n+\ell+1)a_o}. \tag{65}$$

Reinstating $\hbar$ and m, we obtain:

PROPOSITION 7

For the radial Nelson diffusion corresponding to above radial wavefunction

$$\gamma_{o,\ell}(a) \sim \frac{8ma_o^2}{\hbar}\left(\frac{\ell+1}{2}\right)^{2\ell+4}(2\ell)!\left(\frac{a_o}{a}\right)^{2\ell+1}. \tag{66}$$

Moreover,

$$\lim_{n\uparrow\infty}\lim_{a\downarrow 0}\frac{\gamma_{n,\ell}(a)}{\gamma_{o,\ell}(a)} = \frac{(2\ell+1)!}{(2\ell+2)(\ell+1)^{2\ell+3}}\frac{1}{8}\int_0^{j_{1,2\ell+1}} x^3 (J_{2\ell+1}(x))^2 dx, \tag{67}$$

$j_{1,2\ell+1}$ being first zero of $J_{2\ell+1}$ Bessel function.

Possible Experimental Tests Revisited

Calculations show that for the first few excited states the above asymptotic results give reasonably good agreement with exact values as calculated by Eq. (52) for $p^+\pi^-$, for instance. Also, one can calculate the numerical value of $\mathbb{E}_{\ell=1}(\tau(a))$ for $p^+\pi^-$ system with lowest possible energy and with angular momentum unity. This gives, as $a \sim \frac{\hbar}{m_\pi c}$,

$$\mathbb{E}_{\ell=1}(\tau(a)) \sim 4.2\times 10^{-12} \text{ secs},$$

which when compared with actual decay rates of 10^{-12} secs looks as if it might be accessible to experiment. Here though one has to guard against being too optimistic as the $P \to S$ transition rate may be too great (see Ref. (4)). Nevertheless stochastic mechanics has an important role to play here in suggesting corresponding results in quantum mechanics.

Batchelor's Computer Tabulation

Zero Angular Momentum Case

n	$\lim_{a\downarrow 0} \gamma^n(a)/\gamma^0(a)$
0	1.00000
1	$.4212\dot{2}$
2	.38735
3	.37686
4	.37221
5	.36973
6	.36826
7	.36731
8	.36666
9	.36620
10	.36586
11	.36560
12	.36540
13	.36524
14	.36511
15	.36500
16	.36492
17	.36484
18	.36478
19	.36473

Note that:

$$\frac{1}{16}\int_0^{j_{1,1}} u^3 (J_1(u))^2 du = .36424 \quad (5 \text{ d.p.})$$

ACKNOWLEDGEMENT

It is a pleasure to thank Professors John Taylor and David Williams at Cambridge for helpful conversations.

REFERENCES

[1] Albeverio, S., Blanchard, Ph., Hoegh-Krohn, R. (1984): Newtonian Diffusions and Planets with a remark on non-standard Dirichlet forms and Polymers. In 'Stochastic Analysis and Applications', Proceedings, Swansea 1983, editors

A. Truman and D. Williams, 1 - 25. Lecture Notes in Maths. 1095, Springer-Verlag.

[2] Batchelor, A., Truman, A., First Hitting Times for Excited States, in preparation.

[3] Batchelor, A., Truman, A., First Hitting Times in Stochastic Mechanics, in preparation.

[4] Bethe, H.A., Leon, M. (1962): Negative Meson Absorption in Liquid Hydrogen. Phys. Rev. 127, 636 - 647.

[5] Mandl, P. (1968): Analytical Treatment of One-Dimensional Markov Processes. Springer-Verlag, New York.

[6] Nelson, E. (1967): Dynamical Theories of Brownian Motion. Mathematical Notes. Princeton: Princeton University Press.

[7] Nelson, E. (1985): Quantum Fluctuations. Princeton Series in Physics. Princeton: Princeton University Press.

[8] Newell, G.F. (1962): Asymptotic Extreme Value Distribution for One-Dimensional Diffusion Processes. J. of Math. and Mech. 11, 481 - 496.

[9] Truman, A. (1986): An Introduction to the stochastic mechanics of stationary states with applications. In 'From local times to global geometry, control and physics', editor K.D. Elworthy, 329 - 344. Pitman Research Notes in Maths. Series 150. Longman Scientific and Technical.

[10] Williams, D., Rogers, L.C.G. (1987): Diffusions, Markov Processes and Martingales, Volume 2 Itô Calculus. Chichester: John Wiley.

LIMIT THEOREMS FOR STOCHASTIC PROCESSES ASSOCIATED WITH A BOSON GAS

M. van den Berg
Department of Mathematics, Heriot-Watt University
Riccarton, Edinburgh EH14 4AS, United Kingdom

J.T. Lewis
Dublin Institute for Advanced Studies
10 Burlington Road, Dublin 4, Republic of Ireland

§1 INTRODUCTION

In this lecture, we discuss the density of particles having energy less than λ in a boson system as a stochastic process indexed by λ. The notation is that of [1] in this volume. Recall that the hamiltonian for the free boson gas is given by

$$H_\ell(\omega) = \sum_{j\geq 1}\lambda_\ell(j)\sigma_j(\omega), \tag{1.1}$$

where $0 = \lambda_\ell(1) \leq \lambda_\ell(2) \leq \ldots$. For a system in a region of volume V_ℓ, the grand canonical pressure $p_\ell(\mu)$ is defined for $\mu < 0$ by

$$p_\ell(\mu) = \frac{1}{\beta V_\ell}\,\ell n\left\{\sum_{\omega\in\Omega} e^{\beta(\mu N(\omega)-H_\ell(\omega))}\right\}. \tag{1.2}$$

In [2] in this volume, we recalled results (proved in [3]) on the existence of the pressure in the thermodynamic limit:

$$p(\mu) = \lim_{\ell\to\infty} p_\ell(\mu). \tag{1.3}$$

In order to discuss the phenomenon of boson condensation, we introduced in [3] the family of random variables $\{X_\ell(\cdot;\lambda):\lambda \geq 0\}$ defined by

$$X_\ell(\omega;\lambda) = \frac{1}{V_\ell}\sum_{\{j:\lambda_\ell(j)\leq\lambda\}}\sigma_j(\omega). \tag{1.4}$$

For the free boson gas, we have the following result:

THEOREM 1

Suppose that (S1) *and* (S2) *hold; then, for* ρ_c *finite,*

$$\lim_{\lambda\downarrow 0}\lim_{\ell\to\infty} \mathbb{E}_\ell^\rho[X_\ell(\lambda)] = (\rho-\rho_c)^+. \tag{1.5}$$

[Conditions (S1) and (S2) and the critical density ρ_c are defined in §2 and §3 of [2] in this volume. Here $\mathbb{E}_\ell^\rho[\cdot]$ denotes the expectation taken with respect to the grand canonical probability measure $\mathbb{P}_\ell^\mu[\cdot]$ with $\mu = \mu_\ell(\rho)$, defined in §3 of

[2]; it is the expectation at fixed mean density ρ.]

Proof:

From the definition of $X_\ell(\lambda)$, we have

$$\mathbb{E}_\ell^\rho[X_\ell(\lambda)] = \int_{[0,\lambda)} p'(\mu_\ell(\rho)|\lambda)dF_\ell(\lambda) = \rho - \int_{[\lambda,\infty)} p'(\mu_\ell(\rho)|\lambda)dF_\ell(\lambda). \tag{1.6}$$

But, for $\mu < \lambda$, the sequence

$$\left\{ \int_{[\lambda,\infty)} p'(\mu|\lambda)dF_\ell(\lambda) : \ell = 1,2,\dots \right\} \tag{1.7}$$

converges uniformly in μ on compacts to

$$\int_{[\lambda,\infty)} p'(\mu|\lambda)dF(\lambda). \tag{1.8}$$

Hence, by Proposition 2 of [2], we have for $\lambda > 0$:

$$\lim_{\ell\to\infty} \mathbb{E}_\ell^\rho[X_\ell(\lambda)] = \rho - \int_{[\lambda,\infty)} p(\mu(\lambda)|\lambda)dF(\lambda). \tag{1.9}$$

But, by hypothesis, ρ_c is finite so that we may invoke the dominated convergence principle to conclude that

$$\lim_{\lambda\downarrow 0} \int_{[\lambda,\infty)} p'(\mu(\rho)|\lambda)dF(\lambda) = \int_{[0,\infty)} p'(\mu(\rho)|\lambda)dF(\lambda) = \begin{cases} \rho \ , & \rho < \rho_c, \\ \rho_c \ , & \rho \geqslant \rho_c. \end{cases}$$

Thus we have

$$\lim_{\ell\to\infty} \mathbb{E}_\ell^\rho[X_\ell(\lambda)] = (\rho - \rho_c)^+ \qquad \blacksquare$$

In the free boson gas there is a second effect, discovered by M. Kac in 1971. We saw in §3 of [2] that the free-energy has a first-order phase-transition segment $[\rho_c,\infty)$; it follows that for $\rho > \rho_c$ there is no guarantee that the weak law of large numbers will hold for the distribution $\mathbb{K}_\ell^\rho = \mathbb{P}_\ell^\rho \circ X_\ell^{-1}$ of the number density $X_\ell = N/V_\ell$. In fact, there is no guarantee that for, $\rho > \rho_c$, the sequence $\{\mathbb{K}_\ell^\rho : \ell = 1,2,\dots\}$ will converge; nevertheless, by the Helly Selection Principle, a subsequence will converge, but the limit distribution will depend on the detailed behaviour of the corresponding subsequence of the sequence $\{\lambda_\ell(\cdot) : \ell = 1,2,\dots\}$. In other words, it is possible to have two sequences, $\{\lambda_\ell(\cdot) : \ell = 1,2,\dots\}$ and $\{\hat{\lambda}_\ell(\cdot) : \ell = 1,2,\dots\}$, each satisfying (S1) and (S2) and having the same integrated density of states $F(\cdot)$ but having limit distributions $\mathbb{K}^\rho$ and $\hat{\mathbb{K}}^\rho$ which are distinct for $\rho > \rho_c$. (For $\rho < \rho_c$, they must both be equal to δ_ρ, the degenerate distribution concentrated at ρ, by Theorem 1 of [1].) For example, Kac showed that in the standard example (described in §3 of [2]) the limit distribution is the exponential distribution supported on $[\rho_c,\infty)$ with mean ρ, for $\rho > \rho_c$; other examples are investigated in detail in [3]. We shall see in the next section that, in the mean-field model, this phenomenon disappears: there is no first-order phase-transition segment, the grand canonical pressure exists for all values of μ and is a differentiable function; the weak law of large numbers holds for X_ℓ for all values of the mean density ρ; nevertheless, condensation persists. In these circumstances it becomes interesting to regard $\lambda \longrightarrow X_\ell(\cdot;\lambda)$ as a stochastic process and to enquire about the convergence in distribution of a re-scaled, centred version of it. This we do in §3.

§2 THE MEAN FIELD-MODEL

To describe the mean-field model, we define a sequence of hamiltonians $\{\tilde{H}_\ell : \ell = 1,2,\ldots\}$ by

$$\tilde{H}_\ell(\omega) = H_\ell(\omega) + \frac{a}{2V_\ell} N^2(\omega) \tag{2.1}$$

with $a > 0$. The term $\frac{a}{2V_\ell}N^2$, which provides a crude caricature of the interaction, can be understood classically: it arises in an "index of refraction" approximation in which we imagine each particle to move through the system as if it were moving in a uniform optical medium and so receiving an increment of energy proportional to the density $X_\ell = N/V_\ell$; since a is positive the interaction is repulsive.

First, we compute the pressure $\tilde{p}_\ell(\mu)$, as explained in §4 of [1]: writing $u(x) = (\mu-\alpha)x - \frac{a}{2}x^2$, a straight-forward manipulation gives

$$\tilde{p}_\ell(\mu) = p_\ell(\alpha) + \frac{1}{\beta V_\ell} \ln \mathbb{E}_\ell^\alpha[e^{\beta V_\ell u(X_\ell)}] = p_\ell(\alpha) + \frac{1}{\beta V_\ell} \ln \int_{[0,\infty)} e^{\beta V_\ell u(x)} \mathbb{K}_\ell^\alpha[dx] \tag{2.2}$$

for each $\alpha < 0$, where $\mathbb{K}_\ell^\alpha = \mathbb{P}_\ell^\alpha \circ X_\ell^{-1}$. But $x \longrightarrow u(x)$ is continuous and bounded above and $\{\mathbb{K}_\ell^\alpha : \ell = 1,2,\ldots\}$ satisfies the Large Deviation Principle with rate-function $I^\alpha(x) = p(\alpha) + f(x) - \alpha x$, by Theorem 1 of [2]. Hence, by Varadhan's First Theorem, $\tilde{p}(\mu) = \lim_{\ell\to\infty} \tilde{p}_\ell(\mu)$ exists and is given by

$$\tilde{p}(\mu) = p(\alpha) + \sup_x\{u(x) - I^\alpha(x)\} = \sup_x\{\mu x - \tilde{f}(x)\}, \tag{2.3}$$

where the mean-field free-energy $\tilde{f}(\cdot)$ is given by

$$\tilde{f}(x) = f(x) + \frac{a}{2}x^2. \tag{2.4}$$

Thus we have proved:

THEOREM 2

Suppose that (S1) *and* (S2) *hold; then the mean-field pressure exists for all real* μ *and is given by*

$$\tilde{p}(\mu) = \sup_x\{\mu x - \tilde{f}(x)\}, \tag{2.5}$$

where $x \longrightarrow \tilde{f}(x)$ *is the mean-field free energy, given by* $\tilde{f}(x) = f(x) + \frac{a}{2}x^2$.

Next, we introduce the mean-field expectation functional $\tilde{\mathbb{E}}_\ell^\mu[\cdot]$ defined by

$$\tilde{\mathbb{E}}_\ell^\mu[\cdot] = \mathbb{E}_\ell^\alpha[\cdot e^{\beta M_\ell}]/\mathbb{E}_\ell^\alpha[e^{\beta M_\ell}], \tag{2.6}$$

and the associated probability measure $\tilde{\mathbb{P}}_\ell^\alpha[\cdot]$, where

$$M_\ell = V_\ell u(X_\ell). \tag{2.7}$$

COROLLARY

The mean-field pressure $\mu \longrightarrow \tilde{p}(\mu)$ *is differentiable for all values of* μ. *The sequence of distribution functions* $\{\tilde{\mathbb{K}}_\ell^\mu = \tilde{\mathbb{P}}_\ell^\mu \circ X_\ell^{-1}\}$ *converges weakly to the degenerate distribution* δ_ρ *concentrated at* $\rho = \tilde{p}'(\mu)$ *and satisfies the Large Deviation Principle with constants* $\{V_\ell\}$ *and rate-function* $\tilde{I}^\mu(x) = \tilde{p}(\mu) + \tilde{f}(x) - \mu x$.

Proof:

Since $x \longrightarrow f(x)$ is strictly convex for $0 \leq x < \rho_c$ and constant for $\rho_c \leq x < \infty$ and $x \longrightarrow \frac{a}{2}x^2$ is strictly convex for $0 \leq x < \infty$, the function $x \longrightarrow \tilde{f}(x) = f(x) + \frac{a}{2}x^2$ is strictly convex for $0 \leq x < \infty$; hence there is no first-order phase-transition segment; equivalently, $\mu \longrightarrow \tilde{p}(\mu)$, the Legendre transform of $x \longrightarrow \tilde{f}(x)$, is differentiable for $\mu < \infty$. It follows from Theorem 1 of [1] that $\tilde{\mathbb{K}}_\ell^\mu \to \delta_\rho$, where $\rho = \tilde{p}'(\mu)$, and from Theorem 4 of [1] that $\{\tilde{\mathbb{K}}_\ell^\mu : \ell = 1,2,\ldots\}$ satisfies the Large Deviation Principle with constants $\{V_\ell\}$ and rate-function $\tilde{I}^\mu(\cdot)$ ■

Although the first-order phase-transition segment, which was present in the free energy function of the free-gas, has disappeared, the phenomenon of condensation persists:

THEOREM 3

Suppose that (S1) *and* (S2) *hold; then, for* ρ_C *finite, we have*

$$\lim_{\lambda \downarrow 0} \lim_{\ell \to \infty} \tilde{\mathbb{E}}_\ell^\mu [X_\ell(\lambda)] = (\rho - \rho_C)^+ , \tag{2.8}$$

where $\tilde{\mathbb{E}}_\ell^\mu[\cdot]$ *is the mean-field expectation functional and* $\rho = \tilde{p}'(\mu)$.

Proof:

First, we remark that an elementary exercise yields the following alternative formula for the mean-field pressure $\tilde{p}(\mu)$:

$$\tilde{p}(\mu) = \inf_{\alpha<0}\left\{\frac{(\mu-\alpha)^2}{2a} + p(\alpha)\right\}, \tag{2.9}$$

where $p(\alpha)$ is the free-gas pressure. The idea of the proof of (2.8) is that we compute the cumulant generating function of $X_\ell(\lambda)$; since

$$V_\ell X_\ell(\lambda) = V_\ell X_\ell - \sum_{\{j:\lambda_\ell(j)>\lambda\}} \sigma_j \tag{2.10}$$

we get

$$\tilde{\mathbb{E}}_\ell^\mu\left[e^{\beta s V_\ell X_\ell(\lambda)}\right] = \tilde{\mathbb{E}}_\ell^{(s,\lambda),\mu}\left[e^{\beta s V_\ell X_\ell}\right],$$

where $\tilde{\mathbb{E}}_\ell^{(s,\lambda),\mu}[\cdot]$ is the mean-field expectation functional for which the free-gas hamiltonian has been modified by the addition of the term $\sum_{\{j:\lambda_\ell(j)>\lambda\}} s\sigma_j$. These considerations yield the formula

$$\lim_{\ell\to\infty} \tilde{\mathbb{E}}_\ell^\mu[X_\ell(\lambda)] = \frac{\partial}{\partial s}\tilde{p}(\mu+s;s,\lambda)\Big|_{s=0}, \tag{2.11}$$

where

$$\tilde{p}(\mu+s;s,\lambda) = \inf_{\alpha<0}\left\{\frac{(\mu+s-\alpha)^2}{2a} + p(\alpha;s,\lambda)\right\} ,$$

and

$$p(\alpha;s,\lambda) = \int_{[0,\lambda)} p(\alpha|\lambda)dF(\lambda) + \int_{[\lambda,\infty)} p(\alpha|s+\lambda)dF(\lambda).$$

A standard argument, using Griffith's Lemma, yields the result. ■

§3 FLUCTUATIONS IN THE MEAN-FIELD MODEL

Fluctuations in $X_\ell = N/V_\ell$ in the mean-field model in the thermodynamic limit were studied for the standard example, described in §3 of [2] in this volume, by Davies [4], Wreszinski [5], Fannes and Verbeure [6] and Buffet and Pule [7]. The mean-field model in the general situation, where the only assumptions about the single-particle spectrum are that (S1) and (S2) hold, was investigated in [8]; we have summarized the results of [8] in §2 and now go on to investigate the fluctuations in X_ℓ. In fact, we do rather more; we regard $\lambda \longrightarrow X_\ell(\lambda)$ as a stochastic process and prove a central limit theorem:

THEOREM 4

Let $Z_\ell(\lambda) = V_\ell^{1/2}\{X_\ell(\lambda) - \tilde{\mathbb{E}}_\ell^\mu[X_\ell(\lambda)]\}$*; then, for* $\mu < a\rho_c$, $Z_\ell(\lambda) \xrightarrow{(d)} Z(\lambda)$, *where* $Z(\lambda)$ *is gaussian with mean zero and covariance* $\Gamma(\lambda_1,\lambda_2)$ *given by*

$$\Gamma(\lambda_1,\lambda_2) = J^\mu_{\lambda_1\wedge\lambda_2} - \frac{aJ^\mu_{\lambda_1}J^\mu_{\lambda_2}}{1 + aJ^\mu_\infty} , \tag{3.1}$$

where

$$J^\mu_\lambda = \int_{[0,\lambda)} p''(\alpha(\mu)|\lambda)dF(\lambda) , \tag{3.2}$$

and $\alpha(\mu)$ *is the value of* α *at which* $\inf_{\alpha<0}\left\{\frac{(\mu-\alpha)^2}{2a} + p(\alpha)\right\}$ *is attained.*

Sketch of proof:

The result follows from a routine, but somewhat tedious, calculation of

$$\lim_{\ell\to\infty}\tilde{\mathbb{E}}^\mu_\ell\left[e^{\beta(s_1Z_\ell(\lambda_1)+s_2Z_\ell(\lambda_2))}\right]$$

along the lines of the proof of Theorem 3. ■

It is interesting to identify the process $Z(\cdot)$ in terms of a standard process.

THEOREM 5

Let $B(\cdot)$ *be a* $BM(1)$, *a brownian motion in* $\mathbb{R}^1$ *starting at zero; then, for* $\mu < a\rho_c$,

$$Z(\lambda) \overset{(d)}{=} B(J^\mu_\ell) - \frac{aJ^\mu_\lambda}{1 + aJ^\mu_\infty} B(J^\mu_\infty + 1/a). \tag{3.3}$$

Proof:

A routine computation shows that the mean of the right-hand side of (3.3) is zero and the covariance is the same as that of $Z(\cdot)$, given by (3.1). Hence the two gaussian processes are equal in distribution. ■

The process (3.3) is a modification of a time-changed brownian bridge; it never reaches the point at which it is tied-down but, as a increases, that point comes closer to J^{μ}_{∞}. This shows how, as the strength of the interaction increases, the fluctuations in $Z(\infty)$ are damped down.

It is a little more difficult to deal with the case $\mu > a\rho_C$; we introduce

$$W_\ell(\lambda) = Z_\ell(\infty) - Z_\ell(\lambda) \tag{3.4}$$

and prove in analagous fashion:

THEOREM 6

For $\mu > a\rho_C$, $W_\ell(\lambda) \xrightarrow{(d)} W(\lambda)$, *where* $W(\lambda)$ *is a gaussian process with mean zero and covariance* $\Gamma(\lambda_1,\lambda_2)$ *given by*

$$\Gamma(\lambda_1,\lambda_2) = K^{\mu}_{\lambda_1 \vee \lambda_2}, \tag{3.5}$$

where

$$K^{\mu}_{\lambda} = \int_{[\lambda,\infty)} p''(0|\lambda)dF(\lambda). \tag{3.6}$$

In this case,

$$W(\lambda) \overset{(d)}{=} K^{\mu}_{\lambda} B\left[\frac{1}{K^{\mu}_{\ell}}\right]. \tag{3.7}$$

The method by which we discovered the representations may be of some interest. The stochastic differential equation satisfied by a process $(X_t)_{t\geq 0}$ with filtration $(\mathcal{F}_t)$ is discussed by Nelson [9]; see also McGill [10].

Suppose that a process $(X_t,\mathcal{F}_t)$ *satisfies the stochastic differential equation*

$$X_t = X_s + \int_s^t \sigma(u,X_u)dB(u) + \int_s^t \tau(u,X_u)du; \tag{3.8}$$

then

$$\tau(s,X_s) = \lim_{t \downarrow s} \frac{1}{t - s} \mathbb{E}\,[X_t - X_s | \mathcal{F}_s] \tag{3.9}$$

and

$$\sigma^2(s,X_s) = \lim_{t \downarrow s} \frac{1}{t - s} \mathbb{E}\,[X_t - X_s)^2 | \mathcal{F}_s]. \tag{3.10}$$

Assuming that the processes $Z(\lambda), W(\lambda)$ satisfy stochastic differential equations, the corresponding coefficients σ and τ can be computed using (3.9) and (3.10); this is a routine exercise starting from the expressions (3.1) and (3.5) for the covariances since the processes are gaussian. Obvious time-changes then give the stochastic differential equations for a brownian bridge and a brownian motion respectively.

REFERENCES

[1] J.T.Lewis: The Large Deviation Principle in Statistical Mechanics: an Expository Account (in this volume).

[2] M. van den Berg, J.T. Lewis, J.V. Pule: Large Deviations and the Boson Gas, (in this volume).

[3] M. van den Berg, J.T. Lewis and J.V. Pule: A General Theory of Bose-Einstein Condensation, Helv. Phys. Acta, 59, 1271-1288 (1986).

[4] E.B. Davies: The Thermodynamic Limit for an Imperfect Boson Gas, Commun. Math. Phys. 28, 69-86 (1972).

[5] W.F. Wreszinski: Normal Fluctuations in some Mean-Field Models in Quantum Statistical Mechanics, Helv. Phys. Acta, 46, 844-868 (1974).

[6] M. Fannes, A. Verbeure: The Imperfect Boson Gas, J. Math. Phys. 21, 1809-1818 (1980).

[7] E. Buffet, J.V. Pule: Fluctuation Properties of the Imperfect Boson Gas, J. Math. Phys. 24, 1608-1616 (1983).

[8] M. van den Berg, J.T. Lewis, P. de Smedt: Condensation in the Imperfect Boson Gas, J. Stat. Phys., 37, 697-707 (1984).

[9] E. Nelson: Dynamical Theories of Brownian Motion, New Jersey: Princeton University Press, 1967.

[10] P. McGill: Seminaires de Probabilites, XX, LNM 1204, Springer: Heidelberg 1986.

LARGE DEVIATIONS AND THE BOSON GAS

M. van den Berg
Heriot-Watt University, Edinburgh

J.T. Lewis
Dublin Institute for Advanced Studies, Dublin 4

J.V. Pulé
University College, Dublin 4
and
Dublin Institue for Advanced Studies, Dublin 4

§1 Introduction

In this lecture we review some large deviation results for probability distributions associated with the free boson gas and discuss briefly their application to models of an interacting boson gas. In §2 we describe the probabilistic setting; in §3 we review results on the free boson gas which we shall require; in §4, §5, §6 and §7 we summarize large deviation results in increasing order of sophistication; in §8 we sketch some applications.

§2 The Probabilistic Setting

Our ultimate aim is to compute thermodynamic functions for certain models of an interacting boson gas. The physical relevance of these calculations will not be discussed here; we shall concentrate on the probabilistic aspects of the investigation.

The probability space Ω on which the models are defined is the space of terminating sequences of non-negative integers: an element ω of Ω is a sequence

$$\{w(j) \in N : j = 1, 2, \ldots\}$$

satisfying $\sum_{j \geq 1} w(j) < \infty$.

The basic random variables, *the occupation numbers*, are the evaluation maps $\sigma_j : \Omega \to N$ given by

$$\sigma_j(\omega) = \omega(j) \tag{2.1}$$

The sequence $\{H_l : l = 1, 2 \ldots\}$ of free-gas hamiltonians is defined by

$$H_l(\omega) = \sum_{j \geq 1} \lambda_l(j)\sigma_j(\omega), \tag{2.2}$$

where $\{\lambda_l(j) : j = 1, 2, \ldots\}$is an ordered sequence of real numbers associated with a region Λ_l of some Euclidean space R^d:

$$0 = \lambda_l(1) \leq \lambda_l(2) \leq \ldots \tag{2.3}$$

The total number of particles $N(\omega)$ is defined by

$$N(\omega) = \sum_{j \geq 1} \sigma_j(\omega). \tag{2.4}$$

As in §2 of [1], we are in a position to define for $\mu < 0$, the grand canonical measure $P_l^\mu[\,\cdot\,]$ on Ω and the grand canonical pressure $p_l(\mu)$:

$$P_l^\mu[\omega] = \frac{e^{\beta(\mu N(\omega) - H_l(\omega))}}{e^{\beta V_l p_l(\mu)}} \tag{2.5}$$

where

$$p_l(\mu) = (\beta V_l)^{-1} \ln\Big(\sum_{\omega \epsilon \Omega} e^{\beta(\mu N(\omega) - H_l(\omega))}\Big). \tag{2.6}$$

Because of (2.3), both (2.5) and (2.6) hold for all $\mu < 0$. The mean particle number density $\mathbf{E}_l^\mu[X_l]$, where $X_l = N/V_l$ and $\mathbf{E}_l^\mu[\,\cdot\,]$, denotes the expectation with respect to the probability measure $P_l^\mu[\,\cdot\,]$ is given by

$$\mathbf{E}_l^\mu[X_l] = p_l'(\mu). \tag{2.7}$$

Using an identity known to Euler, we have

$$\exp \beta V_l p_l(\mu) = \prod_{j \geq 1} (1 - e^{\beta(\mu - \lambda_l(j))})^{-1}, \tag{2.8}$$

so that we write

$$p_l(\mu) = V_l^{-1} \sum_{j \geq 1} p(\mu | \lambda_l(j)). \tag{2.9}$$

where the partial pressure $p(\mu|\lambda)$ is given by

$$p(\mu|\lambda) = \beta^{-1} \ln(1 - e^{\beta(\mu - \lambda)})^{-1}. \tag{2.10}$$

Lemma 1. *For each $\mu < 0$, the occupation numbers are independent, geometrically distributed random variables:*

$$P_l^\mu[\sigma_j \geq m] = e^{m\beta(\mu-\lambda_l(j))}. \tag{2.11}$$

Proof: For $\alpha_j \leq 0,\ j = 1, 2, \ldots$, we have

$$\mathbf{E}_l^\mu[e^{\beta(\sum_{j\leq 1}\alpha_j\sigma_j)}] = \prod_{j\geq 1} \frac{(1 - e^{\beta(\mu-\lambda_l(j))})}{(1 - e^{\beta(\mu+\alpha_j-\lambda_l(j))})}.$$

□

It is convenient to introduce the distribution function

$$F_l(\lambda) = (V_l)^{-1}\sharp\{j : \lambda_l(j) \leq \lambda\}; \tag{2.12}$$

with respect to this, (2.9) can be rewritten as

$$p_l(\mu) = \int_{[0,\infty)} p(\mu|\lambda)dF_l(\lambda); \tag{2.13}$$

the mean particle density is given by

$$\mathbf{E}_l^\mu[X_l] = \int_{[0,\infty)} p'(\mu|\lambda)dF_l(\lambda). \tag{2.14}$$

We note that, for each l, $\mu \mapsto p_l(\mu)$ is a convex function defined on $(-\infty, 0)$; we define

$$p_l(0) = \lim_{\mu\uparrow 0} p_l(\mu) = +\infty \tag{2.15}$$

and

$$p_l(\mu) = +\infty, \quad \mu > 0. \tag{2.16}$$

Then each p_l is a closed convex function defined on the whole of R; its essential domain is

$$\text{dom } p_l = (-\infty, 0).$$

In order to prove the existence of the pressure in the thermodynamic limit, it is necessary to make some assumptions about the $\lambda_l(j)$ and the V_l; putting $\phi_l(\beta) = \int_{[0,\infty)} e^{-\beta\lambda}dF_l(\lambda)$, we formulate conditions:

(S1)

$$\phi(\beta) = \lim_{l\to\infty} \phi_l(\beta)$$

exists for all β in $(0,\infty)$

(S2) *$\phi(\beta)$ is non-zero for at least one value of $\beta \in (0,\infty)$.*

These conditions are weak restrictions on the sequences; their verification in a particular instance can involve some hard analysis.

§3 Results Concerning the Free Boson Gas

In this section we review some results on the general theory of the free boson gas; the proofs can be found in [2].

Proposition l. *Suppose that (S1) and (S2) hold; then the following limits exist.*

(1) $$p(\mu) = \lim_{l\to\infty} p_l(\mu), \qquad \mu < 0,$$

(2) $$F(\lambda) = \lim_{l\to\infty} F_l(\lambda).$$

They are related by

$$p(\mu) = \int_{[0,\infty)} p(\mu|\lambda)dF(\lambda).$$

Moreover, we have

$$p'(\mu) = \int_{[0,\infty)} p'(\mu|\lambda)dF(\lambda).$$

The standard example is the following one: let $h_l = -\frac{1}{2}\Delta$ in Λ_l with Dirichlet conditions on $\partial\Lambda_l$ where $\{\Lambda_l : l = 1, 2...\}$ is a sequence of dilations of a convex region in R^d which eventually fills out the whole of R^d; let $\varepsilon_l(1) = \varepsilon_l(2) \leq ...$ be the eigenvalues of h_l and put $\lambda_l(j) = \varepsilon_l(j) - \varepsilon_l(1)$; then $(S1)$ and $(S2)$ hold and $F(\lambda) = C_d\lambda^{d/2}$.

Next we define the critical density ρ_c:

if $\lambda \to p'(0\ |\lambda)$*is integrable on* $[0, \infty)$ *with respect to* F, *put*

$$\rho_c = \int_{[0,\infty)} p'(0\ |\lambda)dF(\lambda); \tag{3.1}$$

put $\rho_c = \infty$ *otherwise.*

It follows from the dominated convergence principle that if ρ_c is finite then

$$\rho_c = \lim_{\mu\uparrow 0} \int_{[0,\infty)} p'(\mu|\lambda)dF(\lambda) = \lim_{\epsilon\downarrow 0} \int_{[\epsilon,\infty)} p'(0|\lambda)dF(\lambda). \tag{3.2}$$

Clearly, if $F(\lambda) \sim \lambda^\sigma$ with $\sigma > 1$ then ρ_c is finite; if ρ_c is finite then $F(\lambda) \to 0$ as $F \downarrow 0$. (In fact, we have the more precise estimate: for $\varepsilon > 0, F(\varepsilon) < \beta\varepsilon e^{\beta\varepsilon}\rho_c$). Note that in the standard example, ρ_c is finite if and only if $d > 2$.

Again it is convenient to follow the standard conventions for convex functions in extending p to the whole of R: we define $p(0)$ by $p(0) = \lim_{\mu\uparrow 0} p(\mu)$ and put $p(\mu) = +\infty, \quad \mu > 0$. Since p is convex and differentiable for $\mu < 0$, $p'_-(0) = \lim_{\mu\uparrow 0} = \rho_c$. Define

$p'_+(0)$ to be $+\infty$ and $p'_-(\mu) = p'_+(\mu) = +\infty$ for $\mu > 0$. Then p is a closed convex function on the whole of R.

The sub-differential ∂p is given by

$$(\partial p)(\mu) = \begin{cases} p'(\mu), & \mu < 0; \\ [\rho_c, \infty), & \mu = 0. \end{cases} \tag{3.3}$$

For fixed l, the function $\mu \mapsto p'_l(\mu)$ is strictly increasing on $(-\infty, 0)$ and $p'_l(\mu) \to 0$ as $\mu \to -\infty$ while $p'_l(\mu) \to \infty$ as $\mu \to 0$ since $\lambda_l(1) = 0$. It follows that the equation

$$p'_l(\mu) = \rho \tag{3.4}$$

has a unique solution $\mu_l(\rho)$ in $(-\infty, 0)$, for each ρ in $(0, \infty)$. On the other hand, for $\rho_c < \infty$, the function $\mu \mapsto p'(\mu)$ increases from zero to ρ_c as μ ranges through $(-\infty, 0)$. It is convenient to define $\mu(\rho)$ for ρ in $(0, \infty)$ to be the unique root of

$$p'(\mu) = \rho \tag{3.5}$$

if $\rho < \rho_c$ and to be zero if $\rho \geq \rho_c$.

Defining

$$\pi_l(\rho) = (p_l \circ \mu_l)(\rho),$$

so that $\pi_l(\rho)$ is the pressure at mean density ρ and $= (p \circ \mu)(\rho)$, we have

Proposition 2. *Suppose that* $(S1)$ *and* $(S2)$ *hold; then*

(1)
$$\lim_{l \to \infty} \mu_l(\rho) = \mu(\rho),$$

(2)
$$\lim_{l \to \infty} \pi_l(\rho) = \pi(\rho),$$

(3)
$$f(x) \equiv \sup(\mu x - \rho(\mu)) = x\mu(x) - \pi(x).$$

Thus we have a first-order phase-transition when ρ_c is finite; the first-order phase-transition segment is $[\rho_c, \infty)$.

§4 Large Deviations of the Particle Number Density

Let $K_l^\mu = P_l^\mu \circ X_l^{-1}$ be the distribution function of the particle number density $X_l = N/V_l$. It follows from Theorem 1 of [1] that, for $\mu < 0$, $\{K_l^\mu\}$ converges weakly to the degenerate distribution δ_ρ concentrated at $\rho = p'(\mu)$. It follows from Theorem 2 of [1] that the Large Deviation upperbound (LD3) holds for $\mu < 0$ with rate-function $I^\mu(\,\cdot\,)$ given by

$$I^\mu(x) = p(\mu) + f(x) - \mu x. \tag{4.1}$$

However, the existence of the pressure is not sufficient to ensure that the Large Deviation lowerbound holds for an arbitrary open subset of $[0, \infty)$ when ρ_c is finite; although ran $\partial p = [0, \infty)$, the existence of the first-order phase-transition segment $[\rho_c, \infty)$ prevents an application of Theorem 3 of [1] to the whole of $[0, \infty)$. Nevertheless, as we shall see, special features of the free boson gas enable establish the Large Deviation lowerbound (LD4).

Theorem 1. *Suppose that (S1) and (S2) hold; then, for $\mu < 0$, the sequence*

$$\{I_l^\mu = P_l^\mu \circ X_l^{-1} : l = 1, 2, ...\}$$

satisfies the Large Deviation Principle with constants $\{V_l : l = 1, 2, ...\}$ and rate function $I^\mu(\,\cdot\,)$ given by

$$I^\mu(x) = \begin{cases} p(\mu) + f(x) - \mu x, & x \geq 0, \\ \infty, & x < 0. \end{cases} \tag{4.2}$$

Proof:

It was proved in §9 of [1] in this volume that (LD1), (LD2) hold and in §6 that (LD3) holds; it remains to prove that, for each open subset G of $[0, \infty)$:

$$\liminf_{l \to \infty} \frac{1}{\beta V_l} \ln K_l^\mu[G] \geq -\inf_G I^\mu(x). \tag{4.3}$$

Let y be an arbitrary point of G; choose $\delta > 0$ so that $B_y^\delta = (y - \delta, y + \delta) \subset G$ and t_l such that $p_l'(\mu + t_l) = y$. Then, as in §8 of [1], we have

$$K_l^\mu[G] \geq e^{\beta V_l \{p_l(\mu + t_l) - p_l(\mu) - t_l y - \delta |t_l|\}} K_l^{\mu + t_l}[B_y^\delta].$$

By Proposition 2, $\mu + t_l \to \mu(y)$ and $p_l(\mu + t_l) \to p(\mu(y))$ so that

$$\liminf_{l \to \infty} \frac{1}{\beta V_l} \ln K_l^\mu[G] \geq p(\mu(y)) - p(\mu) - (\mu(y) - \mu) y$$

$$-\delta |\mu(y) - \mu| + \liminf_{l \to \infty} \frac{1}{\beta V_l} ln K_l^{\mu + t_l}[B_y^\delta].$$

We now have to distinguish two cases: if $y < \rho_c$ then $\mu(y) < 0$ and we make use of the fact that $\mu \mapsto p(\mu)$ is differentiable for $\mu < 0$; then, by Theorem 1 of [1], for all l sufficiently large $K_l^{\mu + t_l}[B_y^\delta] \geq \frac{1}{2}$ and hence

$$\liminf_{l \to \infty} \frac{1}{\beta V_l} \ln K_l^{\mu + t_l}[B_y^\delta] = 0; \tag{4.4}$$

on the other hand, if $y \geq \rho_c$ we have $\mu(y) = 0$ and we must proceed differently.

Lemma 2. *Let N_1 and N_2 be independent non-negative integer-valued random variables with means m_1 and m_2 respectively. Suppose that N_1 is geometrically distributed and that $\delta \geq 1$; then*

$$P[N_1 + N_2 \in B^{\delta}_{m_1+m_2}] \geq \frac{1}{m_1+m_2}\left(\frac{m_1}{m_1+1}\right)^{m_1+m_2+2}. \tag{4.5}$$

Proof:

The interval $B^1_{m_1+m_2} = (m_1 + m_2 - 1, m_1 + m_2 + 1)$ contains a unique integer $n_0 \geq m_1 + m_2$. Now

$$P[N_1 + N_2 \in \beta^{\delta}_{m_1+m_2}] = \sum_{m \in B^{\delta}_{m_1+m_2}} \sum_{n=0}^{m} P[N_1 = m - n]P[N_2 = n]$$

$$\geq \sum_{n=0}^{n_0} P[N_1 = n_0 - n]P[n_2 = n]. \tag{4.6}$$

Since N_1 is geometrially distributed, $n \mapsto P[N_1 = n]$ is a decreasing function so that

$$\sum_{n=0}^{n_0} P[N_1 = n_0 - n]P[N_2 = n] \geq P[N_1 = n_0]P[N_2 \leq n_0] \tag{4.7}$$

Now

$$\begin{aligned} P[N_1 = n_0] &= \frac{1}{m_1+1}\left(\frac{m_1}{m_1+1}\right)^{n_0} \\ &\geq \frac{1}{m_1+1}\left(\frac{m_1}{m_1+1}\right)^{m_1+m_2+1} \end{aligned} \tag{4.8}$$

and

$$P[N_2 \leq n_0] \geq P[N_2 \leq m_1 + m_2] \geq \frac{m_1}{m_1+m_2} \tag{4.9}$$

by Markov's Inequality. Hence

$$P[N_1 + N_2 \in B^{\delta}_{m_1+m_2}] \geq \frac{1}{m_1+m_2}\left(\frac{m_1}{m_1+1}\right)^{m_1+m_2+2}. \tag{4.10}$$

□

Returning to the proof of Theorem 1, it follows from Lemma 1 that σ_1 is geometrically distributed; applying Lemma 2 with $N_1 = \sigma_1$ and $N_2 = N - \sigma_1$, we have $\frac{m_1}{m_1+1} = e^{\beta(\mu+t_l)}$ and $m_1 + m_2 = V_l y$; thus

$$K_l^{\mu+t_l}[B^{\delta}_y] \geq \frac{1}{V_l y} e^{\beta(\mu+t_l)(V_l y+2)} \tag{4.11}$$

for $V_l \geq \frac{1}{\delta}$. It follows that

$$\liminf_{1\to\infty} \frac{l}{\beta V_l} \ln K_l^{\mu+t_l}[B_y^\delta] = 0 \tag{4.12}$$

since, for $y \geq \rho_c$, $\mu + t_l \to 0$. Thus we have, in both cases,

$$\begin{aligned}\liminf_{1\to\infty} \frac{l}{\beta V_l} \ln K_l^\mu[G] &\geq -p(\mu) - f(y) + \mu y \\ &= -I^\mu(y)\end{aligned} \tag{4.13}$$

for all y in G, since δ was arbitrary. Hence

$$\begin{aligned}\liminf_{l\to\infty} \frac{1}{\beta V_l} \ln\ K_l^\mu[G] &\geq \sup_G(-I^\mu(y)) \\ &= -\inf_G I^\mu(y)\end{aligned} \tag{4.14}$$

□

§5 The Large Deviations of a Vector-valued Random Variable.

The Large Deviation result established in §4 enables us to apply Varadhan's Theorems to suitable functions of $X_l = N/V_l$; to deal with functions of the $m+1$ variables $\sigma_1/V_l, \ldots, \sigma_m/V_l,\ N/V_l$ we prove a Large Deviation result for the sequence of probability distributions of a vector-valued random variable.

Define the vector-valued random variable $X_l : \Omega \to R^{m+1}$ by

$$\begin{aligned} X_l^{(1)}(\omega) &= V_l^{-1}\sigma_1(\omega), \\ &\vdots \\ X_l^{(m+1)}(\omega) &= V_l^{-1} \sum_{j>m} \sigma_j(\omega). \end{aligned}$$

In order to prove a Large Deviation result for $K_l^\mu = P_l^\mu o X_l^{-1}$, it is necessary to make a further hypothesis about the single-particle spectrum. First, we define the cumulant generating function $C_l^\mu[\,\cdot\,]$ by

$$C_l^\mu[t] = \frac{1}{\beta V_l} \ln \mathbf{E}_l^\mu[e^{\beta V_l < t, X_l >}]. \tag{5.1}$$

Lemma 3. *Suppose that (S 1) and (S 2) hold and that* $\lim_{l\to\infty} \lambda_l(j) = \lambda(j)$ *exists for* $j = 1\cdots, m+1$*; then the cumulant generating function*

$$C^\mu[t] = \lim_{l\to\infty} C_l^\mu[t]$$

exists for all t in R^{m+1} and is given by

$$C^{\mu}[t] = \begin{cases} p(\mu + t_{m+1} - \lambda(m+1) - p(\mu), & t \in \mathcal{D}_{\mu}; \\ \infty, & \textit{otherwise}. \end{cases} \quad .(5.2)$$

where

$$\mathcal{D}_{\mu} = \{t : t_j + \lambda(j) < -\alpha, j = 1 \cdots, m+1\} \quad (5.3)$$

Proof:

Put

$$p_l^{(j)}(\mu) = \frac{-1}{\beta V_l} \ln(1 - e^{\beta(\mu - \lambda_l(k))}) \quad (5.4)$$

for $1 \leq j \leq m$ and put

$$p_l^{(m+1)}(\mu) = \frac{-1}{\beta V_l} \sum_{j>m} \ln(1 - e^{\beta(\mu - \lambda_l(j))}). \quad (5.5)$$

Since $\lambda_l(j) \to \lambda(j)$ as $l \to \infty$, $p_l^{(j)}(\mu + t_k)$ is defined, for all l sufficiently large, for $\mu < \lambda(j) - t_j$. On the set $\mathcal{D}_{\mu}$, we have, by Proposition 1,

$$\lim_{l \to \infty} p_l^{(m+1)}(\mu + t_{m+1}) \to p(\mu + t_{m+1} - \lambda(m+1)), \quad (5.6)$$

while for $l \leq j \leq m$,

$$\lim_{l \to \infty} p_l^{(j)}(\mu + t_j) \to 0. \quad (5.7)$$

It follows that

$$\lim_{l \to \infty} C_l^{\mu}[t] = p(\mu + t_{m+1} \quad - \quad \lambda(m+1)) - p(\mu) \quad , \quad t \in \mathcal{D}_{\mu}; \quad (5.8)$$

put $C^{\mu}[t] = \infty$ for t in the complement of $\mathcal{D}_{\mu}$. Then $t \mapsto C^{\mu}[t]$ is a closed proper convex function on R^{m+1} with dom $C^{\mu} = \mathcal{D}_{\mu}$; put

$$I^{\mu}[x] = \sup_{t \in R^{m+1}} \{< x, t > -C^{\mu}[t]\}. \quad (5.9)$$

Theorem 2. *Suppose that (S1) and (S2) hold and that $\lim_{l \to \infty} \lambda_l(j) = \lambda(j)$ exists for $l \leq j \leq m+1$; then, for $\mu < 0$, the sequence*

$$\{K_l^{\mu} = P_l^{\mu} \circ X_l^{-1} : l = 1, 2, \cdots\}$$

satisfies the Large Deviation Principle with constants $\{V_l : l = 1, 2, \cdots\}$ and rate-function $I^{\mu}[\,\cdot\,]$.

Proof.

The proof that (LD1) and (LD2) holds follows, as in §9 of [1], by the fact that $I^{\mu}[\,\cdot\,]$ is the Legendre transform of C^{μ} $[\,\cdot\,]$. to prove that (LD3) holds, we follow Ellis [3] and adapt to our situation Gartners's Lemma:

Let K be a non-empty closed subset of R^{m+1} define $I_\mu[K] = \inf_K I^\mu[x]$. If $0 < I^\mu[K] < \infty$ then there exists a finite set $\tau^{(1)}, \ldots, \tau^{(r)}$ of non-zero vectors in R^{m+1} such that, for $c = I^\mu[K] - \epsilon$, $\epsilon > 0$,

$$K \subset \cup_{j=1}^{r} H_+^\mu(\tau^{(j)}; c), \tag{5.10}$$

where $H_+^\mu(\tau; c) = \{x :< x, \tau > -C^\mu[t] \geq c\}$ if $I^\mu[K] = +\infty$ then, for each $R > 0$, there exists a finite set $\tau^{(j)}, \cdots, \tau^{(r)}$ of non-zero vectors in R such that

$$K \subset \cup_{j=1}^{r} H_+^\mu(\tau^{(j)}; R). \tag{5.11}$$

First suppose that K is such that $0 < I^\mu[K] < \infty$; then

$$\begin{aligned} K_l^\mu[K] &\leq \sum_{j=1}^{r} K_l^\mu[H_+^\mu(\tau^{(j)}; c)] \\ &= \sum_{j=1}^{r} K_l^\mu[\{x :< x, \tau^{(j)} > \quad \geq \quad c^\mu[\tau^{(j)}] + c\}]. \end{aligned} \tag{5.12}$$

But by Markov's Inequality,

$$\begin{aligned} K_l^\mu[\{x :< x, \tau^{(j)} > \geq C^\mu[\tau^{(j)}] + c\}] &\leq e^{-\beta V_l\{c^\mu[\tau^{(j)}]+c\}} \int_{R^{m+1}} e^{\beta V_l < x, \tau^{(j)} >} K_l^\mu[dx] \\ &= e^{-\beta V_l\{C^\mu[\tau^{(j)}]+c-c_l^\alpha[\tau^{(j)}]\}} \end{aligned} \tag{5.13}$$

hence

$$\limsup_{l \to \infty} \frac{1}{\beta V_l} \ln K_l^\mu[K] \leq -I^\mu[K] \tag{5.14}$$

since $C_l^\mu[t] \to C^\mu[t]$ and $\epsilon > 0$ was arbitrary. Now suppose that $I^\mu[K] = +\infty$; then

$$\limsup_{l \to \infty} \frac{1}{\beta V_l} \ln K_l^\mu[K] \leq -R$$

for each $R > 0$ and the result follows. To prove that (LD4) holds for an arbitrary open set G of R^{m+1}, let y be an arbitrary point of G and choose $\delta > 0$ such that

$$\prod_{j=1}^{m+1} (y_j - \delta, y_j + \delta) \subset G. \tag{5.15}$$

Then

$$K_l^\mu[G] \geq \prod_{j=1}^{m+1} K_l^{(j),\mu}[(y_j - \delta, y_j + \delta)] \tag{5.16}$$

where $K_l^{(j),\mu}$ is determined by

$$\int_{[0,\infty)} e^{\beta V_l t_j x} K_l^{(j),\mu}[dx] = e^{\beta V_l \{p_l^{(j)}(\mu+t_j) - p_l^{(j)}(\mu)\}}. \tag{5.17}$$

Now

$$\liminf_{l\to\infty} \frac{1}{\beta V_l} \ln K_l^{(m+1),\mu}[(y_{m+1}-\delta, y_{m+1}+\delta)] \geq -I^{(m+1),\mu}(y_o), \tag{5.18}$$

where

$$I^{(m+1),\mu}(x) = \begin{cases} p(\mu) + f(x) - (\mu - \lambda(m+1))x, & x \geq 0; \\ \infty, & x < 0, \end{cases} \tag{5.19}$$

by the reasoning which established Theorem 1. For $1 \leq j \leq m$,

$$\liminf_{l\to\infty} \frac{1}{\beta V_l} \ln K_l^{(j),\mu}[(y_j - \delta, y_j + \delta)] \geq -I^{(j),\mu}(y_j) \tag{5.20}$$

by direct calculation, where

$$I^{(j),\mu}(x) = \begin{cases} -(\mu - \lambda(j))x, & x \geq 0, \\ \infty, & x < 0, \end{cases} \tag{5.21}$$

since σ_j is geometrically distributed (Lemma 1). Hence

$$\liminf_{l\to\infty} \frac{1}{\beta V_l} \ln K_l^{\mu}[G] \geq -\sum_{j=1}^{m+1} I^{(j),\mu}(y_j) = -I^{\mu}(y) \tag{5.22}$$

and since y was an arbitrary point of G

$$\liminf_{l\to\infty} \frac{1}{\beta V_l} \ln K_l^{\mu}[G] \geq \sup_G(-I^{\mu}(y)) = -\inf_G I^{\mu}(y). \tag{5.23}$$

□

§6 A Large Deviation Result for a Banach Space-valued Random Variable

Let $X_l : \Omega \longrightarrow l_+^1$ be defined by

$$X_l^{(0)}(\omega) = V_l^{-1} N(\omega), \quad X_l^{(j)}(\omega) = V_l^{-1}\sigma_j(\omega), \quad j \geq 1;$$

then$K_l^{\mu} = P_l^{\mu} \circ X_l^{-1}$ is a probability measure on $l_+^1 = \{x_j \geq 0 : \sum_{j\geq 0} x_j < \infty\}$. We regard l_+^1 as the positive cone of the real Banach space l^1 ; equipped with the norm topology, l^1 is a complete separable metric space (a Polish space). However, for our purposes, the weak$\star$-topology on l^1 (the $\sigma(l^1, c_0)$ topology induced by the space c_0 of real sequences converging to zero) is the appropriate one for our purposes. The space l^1 equipped with the $\sigma(l^1, c_o)$ topology is not metrizable; nevertheless, the theory of large deviations is still applicable since the σ-field of Borel subsets of l^1 is the same in both the norm topology and in the $\sigma(l^1, c_0)$topology (see Azencott [4] for a full discussion of this point and Yamasaki [5] for the measure theory).

Notice also that each of the measures K_l^μ is supported on the convex set $\{x \in l_+^1 : x_o = \sum_{j\geq 1} x_j\}$ since $N(\omega) = \sum_{j\geq 1} \sigma_j(\omega)$.

The proofs of the results in this section are more technical and we will not give them here.

Lemma 4. *Suppose that (S1) and (S2) hold and that* $\lim_{l\to\infty} \lambda_l(j) = 0$ *for* $j = 1, 2, \ldots$ *then, for* $\mu < 0$, *we have for each* t *in* c_0

$$C^\mu[t] = \lim_{l\to\infty} C_l^\mu[t] = \begin{cases} p(\mu + t_o) - p(\mu), & t \in \mathcal{D}_\mu, \\ \infty, & \textit{otherwise}, \end{cases} \tag{6.1}$$

where

$$C_l^\mu[t] = \frac{1}{\beta V_l} \ln \mathbf{E}^\mu_l [e^{\beta\{t_o N + t_1\sigma_1 + t_2\sigma_2 + \cdots\}}]$$

and

$$\mathcal{D}_\mu = \{t \in c_o : t_o + \mu < 0, t_o + \sup_{j\geq 1} t_j + \alpha < 0\}. \tag{6.2}$$

Let $I^\mu[x] = \sup_{t \in c_o}\{< x, t > -C^\mu[t]\}$; then a straightforward caculation yields

$$I^\mu[x] = \begin{cases} p(\mu) + f(x_o - \sum_{j\geq 1} x_j) - \mu x_o, & x \in \mathcal{D}_\mu^*, \\ \infty, & \text{otherwise}, \end{cases} \tag{6.3}$$

where

$$\mathcal{D}_\mu^* = \{x \in l_+^1 : x_o \geq \sum_{j\geq 1} x_j\}. \tag{6.4}$$

Theorem 3. *Suppose that (S1) and (S2) hold and that* $\lim_{l\to\infty} \lambda_l(j) = 0$ *for* $j = 1, 2, \ldots$; *then, for* $\mu < 0$, *the sequence*

$$\{K_l^\mu = P_l^\mu \circ X_l^{-1} : l = 1, 2, \ldots\}$$

of probability measures on l_+^1 *satisfies the Large Deviation Principle with constants* $\{V_l\}$ *and rate-function* $I^\mu[\,\cdot\,]$.

§7. A Large Deviation Result for the Occupation Measure

We introduce a measure-valued random variable

$$L_l(\omega; B) = \frac{1}{V_l} \sum_{j \geq 1} \sigma_j(\omega) \delta_{\lambda_l(j)}[B]$$

, where $\delta_\lambda[B] = 1$ if λ is in B and is zero otherwise. Then L_l maps Ω into the space $E = M_b^+(R^+)$ of positive bounded measures on the positive real line. Let $K_l^\mu = P_l^\mu \circ L_l^{-1}$ be the induced probability measure on E; in terms of this we can express the expectation of a functional of L_l as an integral over E. For example,

$$E_l^\mu[e^{-\beta a N^2/2V_l}] = \int_E e^{\beta V_l G(m)} K_l^\mu[dm]$$

where $G(m) = -\dfrac{a}{2}\|m\|^2$ and $\|m\| = \int_{[0,\infty)} m(d\lambda)$. But even in this simplest of examples there is a difficulty in applying Varadhan's Theorem (supposing that we have established a Large Deviation result for $\{K_l^\mu\}$. It is this: in order to prove a Large Deviation result, we have to make use of the weak$\star$-topology on E determined by $C_o(R^+)$, the continuous functions vanishing at infinity; but the function $m \mapsto \|m\|$ is not continuous in this topology and Varadhan's Theorem does not apply. We get around this difficulty as follows: we introduce a *cut-off* T and prove a Large Deviation result for $K_l^\mu = P_l^\mu \circ L_l^{-1}$ where now

$$L_l(\omega; B) = V_l^{-1} \sum_{\{j : \lambda_j(j) \leq T\}} \sigma_j(\omega) \delta_{\lambda_l(j)}[B]; \tag{7.1}$$

then we prove an estimate for $P_l^\mu[X_l^T \geq \delta]$ where

$$X_l^T(\omega) = V_l^{-1} \sum_{\{j : \lambda_l(j) > T\}} \sigma_j(\omega). \tag{7.2}$$

We state these results without proof:

Theorem 4. *Suppose that (S1) and (S2) hold; then, for $\mu < 0$, the sequence $\{K_l^\mu = P_l^\mu \circ L_l^{-1}\}$ of probability measures on $M_b^+([O,T])$ satisfies the Large Deviation Principle with constants $\{V_l\}$ and rate-function $I^\mu[\,\cdot\,]$, given by*

$$I^\mu[m] = \sup_{C([O,T])} \{< m, t > - C^\mu[t]\} \tag{7.3}$$

where

$$C^\mu[t] = \begin{cases} \int_{[O,T]} \{p(\mu + t(\lambda)|\lambda) - p(\mu|x)\} dF(\lambda), & \sup_{[O,T]}\{t(\lambda) - \lambda\} < -\mu, \\ \infty, & \textit{otherwise.} \end{cases} \tag{7.4}$$

Lemma 5. *For $\delta > 0$ and T such that*

$$\int_{T,\infty} p'(\mu|\lambda)dF(\lambda) < \frac{\delta}{2},$$

we have

$$\liminf_{l\to\infty} P_l^\mu[X_l^t \le \delta] \ge 1 - e^{-\delta/2} \tag{7.5}$$

§8 Applications

In this section, we sketch some applications to the statistical mechanics of models of the interacting boson gas. In [6] in this volume, we used Theorem 1 of §4 to prove the existence of the pressure in the mean-field model.

In the same way, Theorem 2 of §5 has been used to prove the existence of the pressure in the Huang-Yang-Luttinger model; details will be found in [7]. Let $H_l(\cdot)$ be the hamiltonian of the free boson gas in the region Λ_l; the m-level H-Y-L model **has** hamiltonian

$$H_l^{(m)}(\omega) = H_l(\omega) + \frac{a}{2V_l}\{2N(\omega)^2 - \sum_{j=1}^{m} \sigma_j(\omega)^2\} \tag{8.1}$$

with $a > 0$. It was introduced in [8] and discussed also by Thouless [9]. Using Varadhan's Theorem and Theorem 2 we can prove

Theorem 5. *Suppose that (S1) and (S2) hold; Then the pressure*

$$p^{(m)}(\mu) = \lim_{l\to\infty} p_l^{(m)}(\mu)$$

in the H-Y-L model with hamiltonian (8.1) exists for all real values of μ and is given by

$$p^{(m)}(\mu) = \sup_{\{0\le x_1 \le x_0\}} \{\mu x_0 - f(x_0 - x_1) - \frac{a}{2}(2x_0^2 - x_1^2) - \lambda^0(x_0 - x_1)\} \tag{8.2}$$

where $\lambda^0 = \inf\{\lambda : F(\lambda) > 0\}$.

Remarks:

(1) The results is independent of m for $m \ge 1$, so that it is reasonable to conjecture that the same result holds for the pressure p_l in the H-Y-L model with hamiltonian $H_l^{(\infty)}$; we hoped to prove this using Theorem 3 of §6, but, so far, technical difficulties have prevented us.

(2) No explicit assumption is made concerning the existence of $\lim_{l\to\infty}\lambda_l(j)$, while the $\lambda(j) = \lim_{l\to\infty}\lambda_l(j)$, $j = 1, \ldots, m+1$, occur explicitly in the statement of Theorem 2. The reason is that $\inf\{\lambda : F(\lambda) > 0\} = 0$ implies that $\lambda(j) = 0$ for $j = 1, 2, \ldots$.

are equal.

The H-Y-L model is a special case of the diagonal model [9] for which the hamiltonian is

$$H_l^D(\omega) = H_l(\omega) + \frac{a}{2V_l}\{2N(\omega)^2 - \sum_{j=1}^{\infty}\sigma_j(\omega)^2\}$$

$$+\frac{1}{2V_l}\sum_{i=1}^{\infty}\sum_{j=1}^{\infty}u(\lambda_l(i), \lambda_l(j))\sigma_i(\omega)\sigma_j(\omega). \tag{8.3}$$

The last two terms in this hamiltonian have different asymptotic behaviour for large l. To understand the effect of each of these two terms we study them separately. Therefore we consider the regularized hamiltonian:

$$H_l^R(\omega) = H_l(\omega) + \frac{1}{2V_l}\sum_{i=1}^{\infty}\sum_{j=1}^{\infty}v(\lambda_l(i), \lambda_l(j))\sigma_i(\omega)\sigma_j(\omega). \tag{8.4}$$

If we assume that $v : R \to R$ is continuous, bounded and positive then we can use Theorem 4 and Lemma 5 of §7 to obtain the following result which is proved in [10].

Theorem 6. *Suppose that (S1)and (S2) hold; then the pressure $p^R(\mu)$ corresponding to the sequence of hamiltonians $\{H_L^R\}$ is given by*

$$p^R(\mu) = \sup_{m\in M_b^+(R+)}\{\mu\|m\| - f^R[m]\} \tag{8.5}$$

where

$$f^R[m] = \int_{[0,\infty)}\lambda m(d\lambda) + \frac{1}{2}\int_{[0,\infty)}m(d\lambda)\int_{[0,\infty)}m(d\lambda')v(\lambda,\lambda')$$

$$-\beta^{-1}\int_{[0,\infty)}s(\rho(\lambda))dF(\lambda) \tag{8.6}$$

and

$$s(x) = (1+x)\ln(l+x) - x\ln x; \tag{8.7}$$

here

$$m(d\lambda) = m_s(d\lambda) + \rho(\lambda)dF(\lambda) \tag{8.8}$$

is the Lebesgue decomposition of m with respect to $dF(\lambda)$.

References

[1] J.T. Lewis: The Large Deviation Principle in Statistical Mechanics : An Expository Account, (this volume).

[2] M. van den Berg, J.T. Lewis and J.V. Pulé: A General Theory of Bose-Einstein Condensation, *Helv. Phys. Acta*, **59**, 1271 -1288 (1986).

[3] R. Ellis: Entropy, Large Deviations and Statistical Mechanics, New York: Springer 1985.

[4] R. Azencott: Grandes deviations et applications, École d' Été de Probabilités de Saint-Flour VIII - 1978, 1 - 176, LNM 774, Berlin :Springer 1980.

[5] Y. Yamasaki: Measures on Infinite Dimensional Spaces, Singapore: World Scientific 1985.

[6] M. van den Berg and J.T. Lewis: Limit Theorems for Stochastic Processes Associated with a Boson Gas, (this volume),

[7] M. van den Berg, J.T. Lewis, J.V. Pulé: The Large Deviation Principle and some models of an interacting Boson Gas, to appear in *Commun. Math. Phys.*

[8] K. Huang, C.N. Yang, J.M. Luttinger: Imperfect Bose gas with hard-sphere interactions, *Phys. Rev.*, **105**, 776 - 784 (1957).

[9] D.J. Thouless: The Quantum Mechanics of Many- Body Systems, New York: Academic Press 1966.

[10] M. van den Berg, J.T. Lewis, J.V. Pulé: (in preparation).

STOCHASTIC MECHANICS OF FREE SCALAR FIELDS

by
Eric A. Carlen*
Princeton University
Princeton NJ 08544 U.S.A.

INTRODUCTION

Here we discuss and develop stochastic mechanics in the treatment of quantum field theory. The subject is in its infancy, so we restrict our attention to free fields, where already difficult and interesting questions arise.

The paper is based on the fact that, just as in the stochastic mechanics of finite particle systems, there is a direct relation between the stochastic mechanical description of a field theory and the Schrödinger representation for the dynamics of the corresponding quantum field theory. Either can be recovered from the other, and this suggests two kinds of investigation.

First, if we have a nice Schrödinger representation for the dynamics of a quantum field theory, we can enquire after the sample path properties of the corresponding diffusions in stochastic mechanics. Of course this is the case with the free scalar fields. A very interesting question arising in this direction is how the diffusions corresponding to the single particle states in quantum field theory tend in the non relativistic limit to the familiar free particle diffusions in stochastic mechanics. We will return to this question, and others, later. The questions arising in this kind of investigation seem to be either very easy or very hard, and for the most part we will only answer easy questions here.

Second, and perhaps more interesting from a physical point of view, we can undertake to directly construct the stochastic mechanics of an interacting field, and then to use the correspondence with the Schrödinger equation to construct a self adjoint Hamiltonian for the corresponding quantum field theory. One may object that in the most interesting case of four space-time dimensions, one expects difficulties with the type of Schrödinger representation that works in lower dimensions. The simple connection between stochastic mechanics and the Schrödinger equation used here may then require modification, or it may just plain break down.

On the other hand, stochastic mechanics provides a probabilistic approach to constructing quantum fields that proceeds in real time—there is no analytic continuation from Minkowski space to Euclidean space involved. This may prove useful on curved space-times where such analytic continuation is impossible.

---

* Research partially supported by an N.S.F. postdoctoral fellowship grant

Unfortunately, the interesting problems arising in this kind of investigation seem very difficult at this time. We therefore stick to the first kind of investigation here. Nelson has discussed some of his ideas regarding the second kind of investigation in his Ascona lectures [2].

Some familiarity with the Nelson's book [1] may be useful in reading what follows, though an aquaintance with diffusion theory and elementary quantum mechanics from any source will probably suffice.

In the first section of the paper we develop the connection between stochastic mechanics and the Schrödinger equation on a Gauss space. This permits us to make the transition to infinitely many degrees of freedom.

The second section is devoted to developing the Schrödinger representation for the scalar free field of mass m in considerable detail. Some of the results here may be of intrinsic interest, though cartainly they are not deep.

Finally the third section presents some results on the stochastic mechanical diffusions corresponding to single particle excitations of the field. The results we discuss all follow fairly easily from the results of the second section. Several harder and more interesting questions are raised but not answered. At the present state of the subject, I must hope that the paper will be interesting not only for the new devlopments within, but also for the exposition of some beautiful ideas of Nelson which still await their full development.

STOCHASTIC MECHANICS ON A GAUSSIAN SPACE

Let us begin by recalling some results about diffusions in $\mathbf{R}^N$making the modifications that follow upon everywhere replacing Lebesgue measure on $\mathbf{R}^N$—which doesn't have an infinte dimensional analog—with a Gauss measure—which does.

Let C be a positive definite operator on $\mathbf{R}^N$. The Gauss measure with covariance C is

$$\rho_C(x) = (detC)^{-1/2}(2\pi)^{-N/2}e^{-\frac{1}{2}(x.C^{-1}x)}.$$

We put

$$A_C = \frac{1}{2}(\nabla - C^{-1}x).\nabla$$

and call this operator the *Ornstein-Uhlenbeck generator with covariance C.* The unique process generated by A_C with $\rho_C(x)dx$ as its invariant measure is called the *ground state process with covariance C.*

Now consider any diffusion $t \mapsto \xi_t$ in $\mathbf{R}^N$generated by

$$\mathcal{G} = A_C + b(x,t).\nabla$$

where b is a smooth bounded time dependent gradient vector field on $\mathbf{R}^N$. Then of course ξ_t is smoothly distributed with respect to $\rho_C(x)dx$. Let $\rho(x,t)$ denote the corresponding density; that is $\rho(x,t)$ is defined by

$$\mathbf{Pr}\{\xi_t \in S\} = \int_S \rho(x,t)\rho_C(x)dx$$

for any Borel subset S of $\mathbf{R}^N$.

To say that $\mathcal{G}$ is the generator of $t \mapsto \xi_t$ means that for any smooth, bounded function f on $\mathbf{R}^N \times \mathbf{R}$,

$$\lim_{h\to 0} \mathbf{E}^{\mathbf{Pr}}(\frac{1}{h})\{f(\xi_{t+h}, t+h) - f(\xi_t, t)|\xi_t\} \stackrel{\text{def}}{=} \mathbf{D}f(\xi_t, t) = (\frac{\partial}{\partial t} + \mathcal{G})f(\xi_t, t).$$

The limit taken with the increment in the other time direction also exists and defines the *backward generator* $\mathcal{G}_*$:

$$\lim_{h\to 0} \mathbf{E}^{\mathbf{Pr}}(\frac{1}{h})\{f(\xi_t, t) - f(\xi_{t-h}, t-h)|\xi_t\} \stackrel{\text{def}}{=} \mathbf{D}_*f(\xi_t, t) = (\frac{\partial}{\partial t} + \mathcal{G}_*)f(\xi_t, t).$$

There is a simple expression giving $\mathcal{G}_*$ in terms of $\mathcal{G}$ and ρ. To read this off the analogous result in [1], just write

$$\mathcal{G} = \frac{1}{2}\Delta + (b(x,t) - \frac{1}{2}C^{-1}x).\nabla.$$

Then it follows from the formulae on page 32 of [1] that

$$\mathcal{G}_* = -\frac{1}{2}\Delta + (b(x,t) - \frac{1}{2}C^{-1}x).\nabla - \nabla log(\rho(x,t)\rho_C(x)).\nabla \quad .$$

But $\nabla log(\rho(x,t)\rho_C(x)) = \nabla log\rho(x,t) - C^{-1}x$, so

$$\mathcal{G}_* = -A_C + (b(x,t) - \nabla log\rho(x,t)).\nabla \quad .$$

The change of reference measure hardly changes the form of the corresponding formula in [1]. Again with hardly any change in form from the analogous definitions in [1] we define the time dependent vector fields u, v and b_* by

$$u = \frac{1}{2}\nabla log\rho, \qquad b = u + v, \qquad b_* = v - u = b - \nabla log\rho.$$

Note that with these definitions,

$$v = \frac{1}{2}(b + b_*) \quad and \quad u = \frac{1}{2}(b_* - b).$$

As usual, we call v the *current* velocity and u the *osmotic* velocity. The operations $\mathbf{D}$ and $\mathbf{D}_*$ are stochastic time differentiations, and combining them in the following way, we have Nelson's definition of the *stochastic* acceleration $a(\xi_t)$:

$$a(\xi_t) = \frac{1}{2}(\mathbf{D}_*\mathbf{D} + \mathbf{D}\mathbf{D}_*)\xi_t.$$

Clearly one computes $\frac{1}{2}(\mathbf{D}_*\mathbf{D} + \mathbf{D}\mathbf{D}_*)$ by evaluating $\frac{1}{2}(\mathcal{G}_*\mathcal{G} + \mathcal{G}\mathcal{G}_*)\xi_t$ at (ξ_t, t). This is easy to do. $\mathcal{G}x = -\frac{1}{2}C^{-1}x + b$ and so $\mathcal{G}_*\mathcal{G}x = \frac{\partial}{\partial t}b - \frac{1}{4}C^{-2}x + \frac{1}{2}C^{-1}b_* - A_C b + b_*.\nabla b$. Averaging this with the analogous computation for $\mathcal{G}\mathcal{G}_* x$ yields:

$$a(\xi_t) + \frac{1}{4}C^{-2}\xi_t = \frac{\partial}{\partial t}v - A_C u + \frac{1}{2}C^{-1}u + v.\nabla v - u.\nabla u \quad .$$

(We have eliminated b and b_* in favor of u and v, and the right hand side is evaluated at (ξ_t, t).)

If we specify $a(\xi_t)$ by equating it to an external force, as in Newtonian mechanics, we have an evolution equation for v in terms of u and the given force. We can get an equation for u from the evolution equation for the density. Adapting formula (5.24) of [1] to our Gaussian reference measure we see that the density ρ satisfies the *continuity equation*:

$$\frac{\partial}{\partial t}\rho = -\nabla.(v\rho) + (C^{-1}x.v)\rho \quad .$$

Since $\frac{\partial}{\partial t}u = \frac{1}{2}\nabla(\rho^{-1}\frac{\partial}{\partial t}\rho)$,

$$\frac{\partial}{\partial t}u = -A_C v + \frac{1}{2}C^{-1}v - v.\nabla u - u.\nabla v \quad .$$

Once a force is specified, say by setting $a(\xi_t) = -\nabla V_{tot}(\xi_t)$, we have a coupled pair of equations for u and v. Once we have solved them (for some chosen initial conditions), we have the *drift field* $b = u+v$ of a diffusion satisfying $\frac{1}{2}(\mathbf{D}_*\mathbf{D}+\mathbf{D}\mathbf{D}_*) = -\nabla V_{tot}(\xi_t)$. Of course it still remains to solve a stochastic differential equation for the diffusion itself.

Nelson discovered that with the force specified as above, it is easy to solve this non-linear coupled system of equations because a change of dependent variable transforms it into the Schrödinger equation.

Suppose V is smooth bounded function on $\mathbf{R}^N$ and ψ satisfies the Schrödinger equation

$$i\frac{\partial}{\partial t}\psi(x,t) = \big(-A_C + V(x)\big)\psi(x,t) \quad .$$

Suppose also $\int |\psi(x,t)|^2 \rho_C(x)dx = 1$ for some, and hence all, t. Define the functions R and S by

$$\psi(x,t) = e^{R(x,t)+iS(x,t)}$$

and, temporarily abusing notation, define the vector fields u and v by $u = \nabla R$ and $v = \nabla S$. Then the real and imaginary parts of the Schrödinger equation yield the following pair of equations for u and v:

$$\frac{\partial}{\partial t}v = A_C u + u.\nabla u - v.\nabla v - \frac{1}{2}C^{-1}u - \nabla V$$

$$\frac{\partial}{\partial t}u = -A_C v - u.\nabla v - v.\nabla u + \frac{1}{2}C^{-1}v \quad .$$

This equation for this u is the same as our previous equation for our previous u. And if we identify

$$a(\xi_t) + \frac{1}{4}C^{-2}\xi_t = -\nabla V(\xi_t) \quad ,$$

the two equations for the two vector fields v coincide. This is the same as putting $a(\xi_t) = -\nabla V_{tot}(\xi_t)$ where $V_{tot}(x) = V(x) + \frac{1}{8}x.C^{-1}x$.

Working in the other direction, we define $R(x,t) = \frac{1}{2}log(\rho(x,t))$ so that $u = \nabla R$. Since by assumption b, and hence v, is a gradient, we define $S(x,t) + \alpha(t)$ by $v = \nabla S$. The notation explicitly takes into account the fact that for each t, S is only defined up to a constant. One then easily sees that if u and v satisfy the equations (1.5) and (1.6), and if $a(\xi_t)$ is defined by (1.9), then a proper choice of S can be made so that ψ defined in terms of R and S by (1.8) satisfies the Schrödinger equation (1.7).

This proves the following result:

Theorem 1.1 A smooth diffusion generated by

$$\mathcal{G} = \frac{\partial}{\partial t} + A_C + b.\nabla$$

with density ρ satisfies Nelson's equation

$$\frac{1}{2}(\mathbf{D}_*\mathbf{D} + \mathbf{D}\mathbf{D}_*)\xi_t = -\frac{1}{4}C^{-2}\xi_t - \nabla V(\xi_t)$$

exactly when there is a solution of the Schrödinger equation

$$i\frac{\partial}{\partial t}\psi(x,t) = \big(-A_C + V(x)\big)\psi(x,t)\xi_t$$

so that

$$b = Re\frac{\nabla\psi}{\psi} + Im\frac{\nabla\psi}{\psi} \quad and \quad \rho(x,t) = |\psi(x,t)|^2 \quad .$$

Now let us consider the infinite dimensional case, first on an informal level. We substitiute $L^2(\mathbf{R}^{d-1}, dx)$ for $\mathbf{R}^N$ and take $C = \frac{1}{2}(-\Delta + m^2)^{-1/2}$. Here $d - 1 \geq 2$ and $m > 0$. Then when $V = 0$, the free case, we have an equivalence between solutions of

$$i\frac{\partial}{\partial t}\Psi = -A_{\frac{1}{2}(-\Delta+m^2)}\Psi$$

and diffusions satisfying

$$(*) \qquad \frac{1}{2}(\mathbf{D}_*\mathbf{D} + \mathbf{D}\mathbf{D}_*)\xi_t = -(-\Delta + m^2)\xi_t \quad .$$

If the paths of $t \mapsto \xi_t$ were smooth, our stochastic time derivatives would coincide with the usual time derivatives, and we would have

$$\frac{1}{2}(\mathbf{D}_*\mathbf{D} + \mathbf{D}\mathbf{D}_*)\xi_t = \left(\frac{\partial}{\partial t}\right)^2 \xi_t$$

and so $t \mapsto \xi_t$ would then satisfy the *Klein-Gordon* equation

$$\left(\left(\frac{\partial}{\partial t}\right)^2 - \Delta + m^2\right)\xi_t = 0 \quad .$$

Of course the paths are not smooth. There is all the same a very interesting connection with the Klein-Gordon equation [2]. First consider the ground state process for this covariance. This does exist as a nice diffusion, although it only has continuous sample paths in the completion of $L^2(\mathbf{R}^{\mathbf{d}-\mathbf{1}}, dx)$ in a weaker norm. It is thoroughly investigated at the sample path level in Carmona's paper [3] in a context we will develop more carefully later in the paper. The results of Albeverio and Høegh-Krohn [4] also provide a nice realization of the ground state process. For now we simply state that it follows from these authors' results and the computations sketched above that if we choose an appropriate norm on the state space, the ground state process does actually solve the stochastic mechanical equations of motion. Now let $\phi(x,t)$ be a smooth solution of the Klein-Gordon equation:

$$\left(\left(\frac{\partial}{\partial t}\right)^2 - \Delta + m^2\right)\phi(x,t) = 0 \quad .$$

Consider $t \mapsto \phi(\cdot, t)$ as a path in the state space, and define a new diffusion $t \mapsto \xi_t^\phi$ by the Cameron-Martin shift

$$\xi_t^\phi = \xi_t + \phi(\cdot, t) \quad .$$

Of course the transformed diffusion has a generator of the form we are discussing, and since the equation $(*)$ is *linear* in ξ_t, and since the deteministic process $t \mapsto \phi(\cdot, t)$ satisfies this equation, so does $t \mapsto \xi_t^\phi$.

So we can produce a large family of solutions to our stochastic mechanical equation of motion just by adding solutions of the Klein-Gordon equation to the ground state process. The corresponding solutions of the Schrödinger equation are the *coherent state* solutions of the Schrödinger equation for the free quantum scalar field of mass m. We will write down these wave functions in the next section after we have carefully formulated the Schrödinger representation for the quantum dynamics of the free scalar field.

Guerra and Ruggerio [9] were the first to investigate the stochastic mechanics of free scalar fields. They put cutoffs on the classical field theory to obtain a system of harmonic oscillators; they then treated these stochastic mechanically. The minimal energy process so obtained is Gaussian with mean zero and an explicitly computable correlation. Removing the cutoffs at the level of the correlation, they identified the minimal energy process as the ground state process of the corresponding Euclidean quantum field theory.

Our approach to the subject avoids cutoffs — which are not needed in the free case. While it provides a slightly more direct route to the result of Guerra and Ruggerio, its real advantage is in treating the processes corresponding to N-particle excitations of the field. It is when looking at these N-particle diffusions that we should expect to see interesting sample path properties. While the sample path properties of the coherent state diffusions are related to the sample path properties of the ground state diffusion in a transparent way, much more interesting—but rather difficult—questions arise in the consideration of the diffusions associated to N-particle excitations of the field. Before we can discuss these, we need to produce the corresponding solutions of the Schrödinger equation; we do this in the next section.

Before leaving this section, it is useful to rewrite some of our results in the form they assume upon changing variables so that our Gauss measure becomes a unit Gauss measure. This is easy to do. Let new variables y be defined by

$$y = C^{-1/2}x$$

so that

$$\rho_C(x)dx = \rho_I(y)dy \quad .$$

Under this change of variables, A_C becomes $\hat{A}_C$ where

$$\hat{A}_C = \frac{1}{2}(\nabla_y - y).C^{-1}\nabla_y \quad .$$

Continuing to use a hat to denote the change of coordinates, we have $\hat{b} = C^{-1/2}b$ and $\hat{b}_* = C^{-1/2}b_*$. Defining $\hat{u}$ and $\hat{v}$ in terms of $\hat{b}$ and $\hat{b}_*$ as before, we get yet another power of $C^{-1/2}$ when we express $\hat{u}$ in terms of $\hat{\rho}$ since $\hat{u}$ is a gradient:

$$\hat{u} = \frac{1}{2}C^{-1}\nabla_y\hat{\rho} \quad .$$

Similarly, the continuity equation becomes

$$\frac{\partial}{\partial t}\hat{\rho} = -\big((\nabla_y.\hat{v}) + 2(C\hat{u}.\hat{v}) - (y.\hat{v})\big)\hat{\rho}$$

and the formula relating stochasic mechanical diffusions to solutions ψ of the Schrödinger equation becomes

$$\hat{b}(y,t) = C^{-1}\left(Re\frac{\nabla_y\psi(y,t)}{\psi(y,t)} + Im\frac{\nabla_y\psi(y,t)}{\psi(y,t)}\right) \quad .$$

Unfortunately, when V is not quadratic, there is no simple way to directly produce a large family of solutions to our stochastic mechanical equations of motion. Without linearity in $(*)$, one must work much harder.

THE SCHRÖDINGER EQUATION FOR THE FREE FIELD

The quantization of the classical scalar field obeying the Klein-Gordon equation

$$((\frac{\partial}{\partial t})^2 - \Delta + m^2)\phi = 0$$

is usually carried out in terms of a second quantized occupation number representation. While this is satisfactory for the computational purposes of orthodox quantum field theory, it is less than satisfactory for our purposes here. We must represent each quantum state with a wavefunction Ψ on a Riemannian manifold $\mathcal{Q}$, the configuration space of our field system. Only then can we develop the connection between wavefunctions on $\mathcal{Q}$ and diffusions in $\mathcal{Q}$ as we did in the last section. For this reason, we must construct the Schrödinger representation for the quantum mechanics of the scalar free field of mass m in a little more detail than usual. This is easy to do, and it may also be useful for other purposes.

First it is important to emphasize the role of the Klein-Gordon equation in this endeavor. The Klein-Gordon equation is not, despite some early confusion, a relativistic generalization of the non-relativistic free Schrödinger equation. Rather, it is a kinematical equation. The Klein-Gordon equation

$$((\frac{\partial}{\partial t})^2 - \Delta + m^2)\phi = 0$$

stands in the same relation to Schrödinger equation

$$i\frac{\partial}{\partial t}\Psi = \mathbf{H}\Psi$$

for the quantized field—we will soon define the terms in this equation—as does the Newton equation for a free particle in $\mathbf{R}^3$

$$(\frac{d}{dt})^2 x(t) = 0$$

to the Schrödinger equation

$$i\frac{\partial}{\partial t}\psi(x,t) = -\frac{1}{2m}\Delta\psi(x,t)$$

for a free particle of mass m in $\mathbf{R}^3$.

In all cases, the kinematical equation is used first to define the classical *phase space* of the system. As a set, this is just the space of initial data for the kinematical equation. Since the particular kinematical equations we are interested in come from Hamiltonian dynamical systems, the initial data space has a natural symplectic structure. With this symplectic structure, it is the classical phase space of the system.

The next step is to identify the phase space, as a symplectic manifold, with the cotangent bundle $T^*(\mathcal{Q})$ of some Riemannian manifold $\mathcal{Q}$, the configuration space of the system. ($\mathcal{Q}$ will be a Hilbert space in the special case we consider here.) The problem with this step—one of the reasons quantization is not an algorithm—is that there may be no unique or even distinguished way to represent the phase space as a cotangent bundle. Even worse, there may be no way to do this consistent with certain symetries of the phase space. This unfortunate circumstance occurs already with the free quantum fields, as we will soon see.

One then chooses a measure on $\mathcal{Q}$. The complex Hilbert space $L^2(\mathcal{Q})$ is the quantum mechanical *state space.* Regarded as a real Hilbert space, it has a natural symplectic structure given by the imaginary part of its inner product. The quantization is finished by writing down a dictionary relating the symplectic structures of $T^*(\mathcal{Q})$ and $L^2(\mathcal{Q})$—it is here that representations of the canonical comutation relations are of interest—and then translating the dynamics from the first setting to the second.

A clear and thorough discussion of quantization at this general level can be found in Segal's book [5]. In our special case, we will be able to translate the dynamics in a fairly transparent way, and so we will not have to worry very much about symplectic structures and representations of the canonical comutation relations. But that is only because we are restricting our attention to free fields here.

We now proceed with the actual quantization of the scalar free field. First, define the *single particle Hamiltonian* H by

$$H = (-\Delta + m^2)^{1/2} \quad .$$

Then $\left(\left(\frac{\partial}{\partial t}\right)^2 - \Delta + m^2\right) = \left(i\frac{\partial}{\partial t} + H\right)\left(-i\frac{\partial}{\partial t} + H\right)$. And so if $\phi(x,t)$ is any smooth solution of the Klein-Gordon equation with compact support in x for fixed t, we can naturally decompose ϕ as

$$\phi(x,t) = \left(\phi_+(x,t) + \phi_-(x,t)\right)$$

where

$$\phi_+(\cdot,t) = e^{-itH}\phi_+(\cdot,0) \quad and \quad \phi_-(\cdot,t) = e^{itH}\phi_-(\cdot,0) \quad .$$

In fact, Fourier transforming in the spatial variables, we have

$$\phi(x,t) = \left(\frac{1}{2\pi}\right)^d \int \left(e^{-it\omega(k)}a_+(k,\omega(k)) + e^{it\omega(k)}a_-(k,\omega(k))\right)\frac{1}{\omega(k)}e^{ik.x}dk$$

where

$$\omega(k) = \sqrt{(k^2+m^2)}$$

and since ϕ is real, $a_- = a_+^*$. Actually, it is easiest to take the Fourier transform in the space and time variables first. Formally one gets a multiple of $\delta(\omega^2 - k^2 - m^2)$, and then since $\delta(\omega^2 - \lambda^2) = \frac{1}{2\lambda}(\delta(\omega+\lambda) + \delta(\omega-\lambda))$, integrating over ω yields the above expression. Clearly then

$$\hat{\phi}_+(k,0) = \left(\frac{1}{2\pi}\right)^d \frac{1}{\omega(k)}a_+(k,\omega(k))$$

so that

$$\int \omega(k)|\hat{\phi}_+(k,0)|^2 dk = \left(\frac{1}{2\pi}\right)^d \int |a_+(k,\omega(k))|^2 \frac{1}{\omega(k)}dk$$

and the right hand side is clearly Lorentz invariant (since $\frac{1}{\omega(k)}dk$ is the invariant measure on $\{(k,\omega) : \omega^2 - k^2 = m^2\}$). Also note that

$$\phi = \frac{1}{2}(\phi_+ + \phi_-) \quad and \quad H^{-1}\dot{\phi} = \frac{1}{2i}(\phi_+ - \phi_-)$$

or

$$\phi = Re\phi_+ \quad and \quad H^{-1}\dot{\phi} = Im\phi_+$$

so we can easily recover the Cauchy data from ϕ_+. This leads to the following result, which is proved in more detail in [5], that if we equip the space of real, smooth, compactly supported (for fixed t) solutions ϕ of the Klein-Gordon equation with the Hilbertian norm

$$|||\phi||| = \|H^{1/2}\phi_+(\cdot,0)\|_{L^2(\mathbf{R}^{d-1})}$$

then the completion is a real Hilbert space $\mathcal{H}_r$, and the natural action of the Poincare group on the smooth solutions extends to an orthogonal representation. The representation is irreducible, but we will not use this fact here.

It is clear from the above that $\mathcal{H}_r$ posseses a natural complex structure, even though we are working with real solutions. Namely,complex multiplication acts on ϕ_+ in the obvious way. *Note however, that this complex structure depends on the choice of the $t = 0$ hyperplane.* We define $\mathcal{H}$ to be this complexification of

$\mathcal{H}_r$. $\mathcal{H}$ is called the *single particle Hilbert space*, and it, equipped with its standard symplectic structure is the phase space of our system.

Note that *time inversion* $C : \mathcal{H} \mapsto \mathcal{H}$ defined by $C\phi(x,t) = \phi(x,-t)$ is a complex conjugation. Let $\mathcal{Q}$ be the eigenspace corresponding to the eigenvalue 1. That is, $\mathcal{Q}$ consists of solutions of the Klein-Gordon equation with $\dot{\phi}(x,0) = 0$. We identify $\mathcal{Q}$ with the real Sobolev space obtained by completing the space of smooth compactly supported functions ϕ in the norm $\|(-\Delta + m^2)^{1/4}\phi\|_{L^2(\mathbf{R}^{d-1})}$. Clearly we may regard $\mathcal{H}$ as $T^*(\mathcal{Q})$ in a natural way, and also $\mathcal{H} = \mathcal{Q} \oplus i\mathcal{Q}$.

Now let **Pr** be the unit Gauss measure on Q. That is, let **B** be a Banach space densely containing $\mathcal{Q}$ so that the norm of **B** is measureable over $\mathcal{Q}$ in the sense of Gross [6]. Let $\mathcal{B}$ be the Borel field of **B**. Then **Pr** is the Borel probability measure on **B** uniquely determined by the finitely additive unit Gauss cylinder measure on $\mathcal{Q}$.

Select such a **B** and put

$$\mathcal{K} = L^2(\mathbf{B}, \mathcal{B}, \mathbf{Pr}) \quad ;$$

that is, $\mathcal{K}$ is complex L^2 of **B**, and loosely speaking, we will say of $\mathcal{Q}$. The fact that **B** is a Banach space is important because it allows us to do differential as well as integral calculus. Other than that, the choice of **B** is quite immaterial. In fact, there is a natural isomorphism between the realizations of $\mathcal{K}$ arising from any two choices of **B**. This is a consequence of the fundamental result of Segal [6] that $\mathcal{K}$ is naturaly isomorphic to the symmetric tensor algebra over $\mathcal{H}$.

We shall soon use this isomorphism in computations. In order to spell it out, we first need to recall the *homogeneous chaos* decomposition of $\mathcal{K}$. $\mathcal{K}_0$ is defined to be the constant subspace of $\mathcal{K}$. $\mathcal{K}_1$ is defined to be the closed linear span of all the random variables of the form

$$(\cdot, \xi)_{\mathcal{Q}}$$

where ξ is in $\mathcal{Q}$. This is only defined pointwise when ξ happens to lie in the dense subspace $\mathbf{B}^*$ of $\mathcal{Q}$, but it is well defined as an element of $L^p(\mathbf{B}, \mathcal{B}, \mathbf{Pr})$ for all finite p by the obvious approximation procedure. And actually, as we have described it, $\mathcal{K}_1$ is already closed. Clearly $\mathcal{K}_0 \perp \mathcal{K}_1$. We define $\mathcal{K}_{(n)}$ to be the closure of all polynomials of degree n or less in the elements of $\mathcal{K}_1$. We then define $\mathcal{K}_n$ to be the orthogonal complement of $\mathcal{K}_{(n-1)}$ in $\mathcal{K}_{(n)}$.

It is then an easy matter to establish that

$$\mathcal{K} = \oplus_{n=0}^{\infty} \mathcal{K}_n \quad .$$

As usual, we use *Wick dots* to denote orthogonal projection from $\mathcal{K}_{(n)}$ to $\mathcal{K}_n$; that is if

$$(\cdot, \xi_1)_{\mathcal{Q}}, \ldots, (\cdot, \xi_n)_{\mathcal{Q}}$$

is a generic monomial in $\mathcal{K}_{(n)}$, we use

$$: (\cdot, \xi_1)_{\mathcal{Q}} \dots (\cdot, \xi_n)_{\mathcal{Q}} :$$

to denote its projection onto $\mathcal{K}_n$.

The effect of the Wick dots can be explicitly expressed using Hermite polynomials. The *nth Hermite polynomial* h_n is defined by

$$h_n(x)e^{-x^2/4} = \left(\frac{x}{2} - \frac{d}{dx}\right)^n e^{-x^2/4} \quad .$$

From the commutation equation

$$\left[\left(\frac{x}{2} + \frac{d}{dx}\right), \left(\frac{x}{2} - \frac{d}{dx}\right)\right] = 1$$

it is immediate that

$$(2\pi)^{-1/2} \int h_n(x)h_m(x)e^{-x^2/2}dx = \delta_{mn}n!$$

and that the leading term of $h_n(x)$ is x^n. Clearly then for any ξ in $\mathcal{Q}$

$$: (\cdot, \xi)^n_{\mathcal{Q}} : \quad = \quad h_n((\cdot, \xi)_{\mathcal{Q}})$$

and the effect of Wick dots on arbitrary polynomials may be computed by polarization and linearity.

Now let $\hat{\mathcal{K}}$ be the symmetric tensor algebra over $\mathcal{H}$. That is,

$$\hat{\mathcal{K}} = \oplus_{n=0}^{\infty} \hat{\mathcal{K}}_n$$

where $\mathcal{K}_0$ is $\mathbf{C}$ and $\hat{\mathcal{K}}_n$ is $\otimes^n_{sym}\mathcal{H}$ with the Hilbertian norm

$$\|(\phi_1 \otimes \dots \otimes \phi_n)_{sym}\|^2_{\hat{\mathcal{K}}_n} = \sum_{\pi \in \Pi_n} (\phi_1, \phi_{\pi 1})_{\mathcal{Q}} \dots (\phi_n, \phi_{\pi N})_{\mathcal{Q}} \quad .$$

Segal's unitary map $S : \hat{\mathcal{K}} \mapsto \mathcal{K}$ is defined on $\hat{\mathcal{K}}_n$ by

$$S((\xi_1 + i\eta_1) \otimes \dots \otimes (\xi_n + i\eta_n))_{sym} =$$

$$: ((\cdot, \xi_1)_{\mathcal{Q}} + i(\cdot, \eta_1)_{\mathcal{Q}}) \dots ((\cdot, \xi_n)_{\mathcal{Q}} + i(\cdot, \eta_n)_{\mathcal{Q}}) : \quad ,$$

and then it is extended by linearity. Here, ξ and η are used to denote generic elements of $\mathcal{Q}$. Selecting the ξ's and the η's from an orthonormal basis for $\mathcal{Q}$ yields a basis for $\hat{\mathcal{K}}$, and the verification that S is an isometry then procedes straight through formula (3.25) above.

We are finaly ready to transfer the dynamics to $\mathcal{K}$. First we lift the time evolution of the Klein-Gordon equation from $\mathcal{H}$ to $\hat{\mathcal{K}}$, and then we move it over to $\mathcal{K}$ with Segal's unitary transformation S.

Let $U(t) : \mathcal{H} \mapsto \mathcal{H}$ be given by $U(t)\phi_+ = e^{-itH}\phi_+$. Define $\Gamma(U(t)) : \hat{\mathcal{K}} \mapsto \hat{\mathcal{K}}$ by

$$\Gamma(U(t))(\phi_1 \otimes \ldots \otimes \phi_n)_{sym} = \big(U(t)\phi_1 \otimes \ldots \otimes U(t)\phi_n\big)_{sym}$$

Finaly define $\mathbf{U}(t) : \mathcal{K} \mapsto \mathcal{K}$ by

$$\mathbf{U}(t) = S\Gamma(U(t))S^{-1} \quad .$$

It is easy to see that $\{\mathbf{U}(t) : t \in \mathbf{R}\}$ is a strongly continuous unitary group on $\mathcal{K}$, so

$$\mathbf{U}(\mathbf{t}) = e^{-it\mathbf{H}}$$

defines a self adjoint operator $\mathbf{H}$ on $\mathcal{K}$. If Ψ_0 is an element of $\mathcal{K}$ and Ψ is defined by $\Psi = e^{-it\mathbf{H}}\Psi_0$, then Ψ satisfies the Schrödinger equation

$$i\frac{\partial}{\partial t}\Psi = \mathbf{H}\Psi \quad .$$

We refer to elements Ψ of K with $\|\Psi\|_{\mathcal{K}} = 1$ as *wave functions.* The above equation describes their time evolution, and we have the usual Born interpretation of the physical meaning of Ψ.

The short route we have taken to writing down a well defined Schrödinger representation for the dynamics of our quantum field proceeded along a natural enough route, but it took advantage of circumstances special to the case of free fields. If we had treated the quantization of the free Klein-Gordon field by means applicable also to interacting fields, we would have written down $\mathbf{H}$ in some detail *before* writing down $\mathbf{U}(t)$. We have avoided some discussion of the canonical commutation relations, but we have a rather abstract definition of $\mathbf{H}$ at this point. This is easily remedied.

If we write $\Gamma(\mathbf{U}(t)) = e^{-it\hat{\mathbf{H}}}$, then it is trivial to write down an explicit expression for $\hat{\mathbf{H}}$ in terms of H. Clearly $\mathbf{H} = S\hat{\mathbf{H}}S^{-1}$. Since we have an explicit exression for S, it is easy to turn this into an explicit expression for $\mathbf{H}$ in terms of H.

It is easiest to compute the effect of $S\hat{\mathbf{H}}S^{-1}$ on elements of $\mathcal{K}_n$ of the form

$$\Psi_\alpha(\xi) = \frac{1}{\sqrt{\alpha_1!\ldots\alpha_m!}}h_{\alpha_1}\big((\xi,\xi_1)_{\mathcal{Q}}\big)\ldots h_{\alpha_m}\big((\xi,\xi_m)_{\mathcal{Q}}\big)$$

where the ξ_j's are taken from an orthonormal basis for $\mathcal{Q}$ all of whose elements lie in the domain of H (on $\mathcal{Q}$). Here of course, α is a multi-index, and $n = \alpha_1 + \ldots + \alpha_m$. Then simply working through the defintions of S and $\hat{\mathbf{H}}$, and using the identities

$$\frac{d}{dx}h_n = nh_{n-1} \quad xh_n = h_{n+1} + nh_{n-1}$$

(which follow immediately from the definition of the Hermite polynomials), one finds

$$\mathbf{H}\Psi_\alpha = \Big(- \sum_{j,k=1}^{\infty} (\xi_j, H\xi_k)_{\mathcal{Q}} \partial_j \partial_k + \sum_{j=1}^{\infty} (\cdot, H\xi_j)_{\mathcal{Q}} \partial_j \Big) \Psi_\alpha \quad .$$

The first order terms in $\mathbf{H}$ serve to replace single factors of $(\xi, \xi_j)_{\mathcal{Q}}$ with a multiple (depending on how many factors there were there) of $(\xi, H\xi_j)_{\mathcal{Q}}$. The second order terms then cancel off the component in $\mathcal{K}_{(n-1)}$ this creates.

Here we use ∂_j to denote the operation of Fréchet differentiation of functions on $\mathbf{B}$ in the direction ξ_j. Later when we write ∇ in front of functions on $\mathbf{B}$, we mean the Fréchet gradient along the subspace $\mathcal{Q}$. For instance, $\nabla\Psi$ is then a $\mathcal{Q}$ valued function on $\mathbf{B}$.

We will need the expicit form of $\mathbf{H}$ in the next section. However for solving the Schrödinger equation it is easier to work directly with $\Gamma(\mathbf{U}(t))$.

Theorem 2.1 The generic solution to

$$i\frac{\partial}{\partial t}\Psi = \mathbf{H}\Psi$$

for $\Psi \in \mathcal{K}_1$ is given by

$$\Psi(\xi, t) = (\xi, \phi(\cdot, t))_{\mathcal{Q}} + i(\xi, H^{-1}\dot{\phi}(\cdot, t))_{\mathcal{Q}}$$

where ϕ is a *finite energy* solution of

$$((\frac{\partial}{\partial t})^2 - \Delta + m^2)\phi = 0 \quad ;$$

that is, a solution with $\phi(\cdot, 0)$ and $H^{-1}\dot{\phi}(\cdot, 0)$ in $\mathcal{Q}$. Ψ is in the form domain of $\mathbf{H}$ on $\mathcal{K}$ precisely when ϕ and $H^{-1}\dot{\phi}$ are in the form domain of H on $\mathcal{Q}$; the analogous statement about operator domains also holds.

Proof The result is an immediate consequence of the previous computations, except perhaps for the statement about form domains. But this follows from the fact that $(\Psi, \mathbf{H}\Psi)_{\mathcal{K}} = \mathcal{E}(\Psi, \Psi)$ for tame functions Ψ where $\mathcal{E}$ is the Dirichlet form

$$\mathcal{E}(\Psi, \Psi) = \sum_{j,k=1}^{\infty} \int_{\mathcal{K}} (\xi_j, H\xi_k)_{\mathcal{Q}} \partial_j \Psi \partial_k \Psi \mathbf{Pr}(d\xi) \quad .$$

In fact it is quite easy to write down an analogous result for solutions in $\mathcal{K}_n$; the only difference is that a few Wick dots crop up. However in the next section we will concentrate on the diffusions corresponding to wave functions in $\mathcal{K}_1$. These are called *single particle* wave functions.

Before leaving the subject of solutions of the free field Schrödinger equation, we briefly discuss the coherent state solutions. We discuss these from the point of view of stochastic mechanics.

Let $t \mapsto \phi(\cdot, t)$ be a finite energy solution of the Klien-Gordon equation. The coherent state diffusion corresponding to ψ is just the ground state process shifted by $\phi(\cdot, t)$ at time t. Therefore its density ρ is just the Radon-Nikodym derivative of the unit Gauss measure shifted by $\phi(\cdot, t)$ with respect to the unit Gauss measure. By the Cameron-Martin formula (in a generality first attained by Segal [6]),

$$\rho(\xi, t) = e^{(\xi,\phi(\cdot,t))_{\mathcal{Q}} - 1/2\|\phi(\cdot,t)\|^2_{\mathcal{Q}}}$$

and since $|\Psi(\xi, t)| = \sqrt{\rho(\xi, t)}$, we need next just find the phase of Ψ. This can be done by extracting v, which is $H\nabla S$, from the continuity equation. Differentiating the above expression for ρ one finds

$$\frac{\partial}{\partial t}\rho(\xi.t) = \big((\xi - \phi(\cdot, t)), \dot{\phi}(\cdot, t)\big)_{\mathcal{Q}} \quad .$$

Since u is the constant (in ξ) vector field

$$u(\xi, t) = H\phi(\cdot t)$$

it follows from the continuity equation that

$$\frac{\partial}{\partial t}\rho(\xi, t) = -(\nabla.v)\rho(\xi, t) + \big((\xi - \phi(\cdot, t)).v\big)\rho(\xi, t) \quad .$$

Simply differentiating our expression for ρ we find

$$\frac{\partial}{\partial t}\rho(\xi, t) = \big((\xi - \phi(\cdot, t)), \phi(\cdot, t)\big)_{\mathcal{Q}}$$

and the equations agree if we have

$$v(\xi, t) = \dot{\phi}(\cdot, t) \quad .$$

This determines S up to a constant; we have

$$S(\xi, t) = \frac{1}{2}(\xi, H^{-1}\dot{\phi}(\cdot, t))_{\mathcal{Q}} + \alpha(t) \quad .$$

In turn, this yields

$$\Psi(\xi, t) = e^{\frac{1}{2}(\xi,\phi(\cdot,t))_{\mathcal{Q}} - \frac{1}{4}\|\phi(\cdot,t)\|^2_{\mathcal{Q}} + \frac{i}{2}(\xi,H^{-1}\dot{\phi}(\cdot,t))_{\mathcal{Q}} + i\alpha(t)}$$

which solves $i\frac{\partial}{\partial t}\Psi = \mathbf{H}\Psi$ provided

$$\dot{\alpha}(t) = \frac{1}{4}(\|H^{1/2}\phi(\cdot,t)\|_{\mathcal{Q}}^2 - \|H{-}1/2\dot{\phi}(\cdot,t)\|_{\mathcal{Q}}^2) \quad .$$

This is solved by

$$\alpha(t) = -\frac{1}{4}(\phi(\cdot,t), H^{-1}\dot{\phi}(\cdot,t))_{\mathcal{Q}} \quad .$$

This finally gives us

$$\Psi(\xi,t) = e^{\frac{1}{2}(\xi,\phi(\cdot,t))_{\mathcal{Q}} - \frac{1}{4}\|\phi(\cdot,t)\|_{\mathcal{Q}}^2 + \frac{i}{2}(\xi,H^{-1}\dot{\phi}(\cdot,t))_{\mathcal{Q}} + \frac{i}{4}(\phi(\cdot,t),H^{-1}\dot{\phi}(\cdot,t))_{\mathcal{Q}}} \quad .$$

It is useful to express this another way. First let us extend the inner product on $\mathcal{Q}$ to be bilinear (*not sesquilinear*) over $\mathbf{C}$ in the obvious way. Then

$$(\phi_+(\cdot,t),\phi_+(\cdot,t))_{\mathcal{Q}} = \|\phi(\cdot,t)\|_{\mathcal{Q}}^2 - \|H^{-1}\dot{\phi}(\cdot,t)\|_{\mathcal{Q}}^2 + 2i(\phi(\cdot,t),H^{-1}\dot{\phi}(\cdot,t))_{\mathcal{Q}}$$

and since $\|\phi(\cdot,t)\|_{\mathcal{Q}}^2 + \|H^{-1}\dot{\phi}(\cdot,t)\|_{\mathcal{Q}}^2$ is a constant of the motion—namely $|||\phi|||^2$—we can rewrite our expression for Ψ as

$$\Psi(\xi,t) = e^{-\frac{1}{8}|||\phi|||^2} e^{\frac{1}{2}(\xi,\phi_+(\cdot,t))_{\mathcal{Q}} - \frac{1}{8}(\phi_+(\cdot,t),\phi_+(\cdot,t))_{\mathcal{Q}}} \quad .$$

Using another Hermite polynomial identity

$$\sum_{n=0}^{\infty} \frac{1}{n!}\lambda^n h_n(x) = e^{(\lambda x - \frac{1}{2}\lambda^2)} \quad ,$$

(the Hermite identities above show that $e^{-x^2/2}\sum_{n=0}^{\infty}\frac{1}{n!}\lambda^n h_n(x)$ and $e^{-(x-\lambda)^2/2}$ both are $L^2(\mathbf{R})$ solutions of the same Sturm-Liouville problem) one sees that

$$\Psi(\cdot,t) = e^{-\frac{1}{8}|||\phi|||^2} S\left(\sum_{n=0}^{\infty} \otimes_{sym}^n \phi_+(\cdot,t)\right) \quad ,$$

which makes it transparent that Ψ satisfies $i\frac{\partial}{\partial t}\Psi = \mathbf{H}\Psi$. Because of the last formula, one usually writes $e^{-\frac{1}{8}|||\phi|||^2} : e^{(\xi,\phi_+(\cdot,t))_{\mathcal{Q}}} :$ for Ψ; this defines the *Wick exponential.*

For some purposes, our first expression for Ψ, though less familiar, is more useful. For any pair (η,π) of elements of $\mathcal{Q}$, define the *coherent state vector* $\Psi_{\eta,\pi}(\xi)$ by

$$\Psi_{\eta,\pi}(\xi) = e^{\frac{1}{2}(\xi,\eta)_{\mathcal{Q}} - \frac{1}{4}\|\eta\|_{\mathcal{Q}}^2 + \frac{i}{2}(\xi,\pi)_{\mathcal{Q}} + \frac{i}{4}(\eta,\pi)_{\mathcal{Q}}} \quad .$$

Let $\{\xi_j : j \in \mathbf{N}\}$ be an orthonormal basis for $\mathcal{Q}$. For any N and any q and p in $\mathbf{R}^n$, let q and p also denote the elements

$$q = q_1\xi_1 + \ldots + q_N\xi_N \quad \textit{and} \quad p = p_1\xi_1 + \ldots + p_N\xi_N \quad .$$

The following small result may be marginally new; however, it is closely related to a well known result about the Weyl transform [10].

Theorem 2.2 Let $\mathcal{F}_N$ be the sigma-algebra $\sigma\{(\cdot, \xi_1)_{\mathcal{Q}}, \ldots, (\cdot, \xi_N)_{\mathcal{Q}}\}$ Then for any Ψ in $\mathcal{K}$, the conditional expectation of Ψ given $\mathcal{F}_N$ may be computed in terms of the coherent states according to the formula

$$E^{\mathbf{Pr}}\{\Phi|\mathcal{F}_N\}(\xi) = (\frac{1}{4\pi})^N \int_{\mathbf{R}^{2N}} dqdp\hat{\Phi}(q,p)\Psi_{p,q}(\xi)$$

where

$$\hat{\Phi}(q,p) = (\Phi, \Psi_{q,p})_{\mathcal{K}}$$

with the inner product being conjugate linear on the right.

Proof For Gaussian spaces orthogonality implies independence. Note then that if $\Phi \perp L^2(\mathbf{B}, \mathcal{F}_N, \mathbf{Pr})$, then both sides of the equation vanish. Therefore, by a simple density argument, it suffices to prove the result when

$$\Phi(\xi) = f((\xi, \xi_1)_{\mathcal{Q}}, \ldots, (\xi, \xi_N)_{\mathcal{Q}})$$

where f is a smooth bounded function on $\mathbf{R}^N$. To show this, we just need to show that

$$(4\pi)^N f(x) =$$

$$\int_{\mathbf{R}^{2N}} \Big(\int_{\mathbf{R}^N} dy (2\pi)^{-N/2} e^{-y^2/2} f(y) e^{\frac{1}{2}(y.q) - \frac{1}{4}|q|^2 - \frac{i}{2}(y.p) - \frac{i}{4}(q.p)}\Big) \times$$

$$e^{\frac{1}{2}(x.q) - \frac{1}{4}|q|^2 + \frac{i}{2}(x.p) + \frac{i}{4}(q.p)} dpdq \quad .$$

But the $(q.p)$ terms cancel out of the exponent, and doing the p integration yields a δ function in $(x - y)$. Doing the y integration, the right hand side is reduced to

$$(2\pi)^{-N/2} \int_{\mathbf{R}^N} dq f(x) e^{-|q-x|^2/2}$$

which is of course just $f(x)$.

This result will not be used in this paper. We present its derivation as a simple example of how stochastic mechanics leads directly to interesting objects in the Schrödinger picture for the free field. However, it may have many uses. Coherent states can be used to provide an approximate diagonalization of perturbations of $\mathbf{H}$ just as in ordinary quantum mechanics one may use coherent states to produce an approximate diagonalization of $-\frac{1}{2}\Delta + V$. Work is underway to determine whether useful spectral bounds may be obtained in this way.

SINGLE PARTICLE DIFFUSIONS

By the results of the first section, the stochastic mechanical diffusion corresponding to a single particle excitation of the scalar free field of mass m satisfies a stochastic differential equation of the form

$$d\xi_t = b(\xi_t, t)dt + dX_t$$

where dX_t is an increment of the ground state process and b is defined by

$$b(\xi_t, t) = u(\xi_t) + v(\xi_t, t) =$$

$$Re\left(\frac{H\nabla\Psi(\xi,t)}{\Psi(\xi,t)}\right) + Im\left(\frac{H\nabla\Psi(\xi,t)}{\Psi(\xi,t)}\right)$$

with $\Psi(\xi, t)$ a single particle solution of the free field Schrödinger equation. Using the results and notation of the last section, we know that any such Ψ is of the form

$$\Psi(\xi, t) = (\xi, \phi_+(\cdot, t))_{\mathcal{Q}} \quad .$$

Since this is a linear function, there is no difficulty computing b. One finds

$$b(\xi,t) = |(\xi, \phi(\cdot,t)_+)_{\mathcal{Q}}|^{-2}\Big(((\xi,\phi(\cdot,t))_{\mathcal{Q}} H\phi(\cdot,t) + (\xi, H^{-1}\dot{\phi}(\cdot,t))_{\mathcal{Q}}\dot{\phi}(\cdot,t))$$

$$+((\xi,\phi(\cdot,t))_{\mathcal{Q}}\dot{\phi}(\cdot,t) - (\xi, H^{-1}\dot{\phi}(\cdot,t))_{\mathcal{Q}} H\phi(\cdot,t))\Big) \quad .$$

Of course, $\rho(\xi, t) = |(\xi, \phi_+(\cdot, t))_{\mathcal{Q}}|^2$.

Fortunately this expression simplifies for large times.

Theorem 3.1 Let $\phi(x,t)$ be a solution of the Klein-Gordon equation with $\phi(\cdot, 0)$ and $H^{-1}\dot{\phi}(\cdot, t)$ in $\mathcal{Q}$. Then

$$\lim_{t\to\infty}(\|\phi(\cdot,t)\|_{\mathcal{Q}}^2 - \|H^{-1}\dot{\phi}(\cdot,t)\|_{\mathcal{Q}}^2) = \lim_{t\to\infty}(\phi(\cdot,t), H^{-1}\dot{\phi}(\cdot,t))_{\mathcal{Q}} = 0 \quad .$$

Proof Fourier transforming,

$$\int (\phi_+(x,t))^2 dx = \int (\hat{\phi}_-(k,t))^* (\hat{\phi}_+(k,t)) dk \quad .$$

Then since for any f, $(\hat{f}(k))^* = \hat{f}^*(-k)$ and as $(\phi_-)^* = \phi_+$,

$$\left(\hat{\phi}_-(k,t)\right)^* = \hat{\phi}_+(-k,t) \quad .$$

So finally

$$\int \left(\phi_+(x,t)\right)^2 dx = \int e^{-2it\sqrt{k^2+m^2}}\hat{\phi}_+(-k,0)\hat{\phi}_+(k,0)dk \quad .$$

Since m is strictly positive, the integrand on the right hand side is in $L^1(\mathbf{R}^{d-1})$, and now a Riemann-Lebesgue argument finishes the proof: First consider any L^1 function f supported outside the ball of radius ϵ about the origin. Let

$$F(t) = \int e^{2it(k^2+m^2)^{1/2}} f(k)dk \quad .$$

Define a change of coordinates Λ_t on $\mathbf{R}^{(d-1)}$ by $\Lambda_t(k) = \lambda_t(k)k$ where λ_t is a positive radial function such that

$$(\Lambda_t(k)^2 + m^2)^{1/2} - (k^2 + m^2)^{1/2} = \pi/2t \quad ;$$

note that $\lambda(k)$ and $|\partial\Lambda_t^{-1}/\partial k|$ tend uniformly to 1 as t increases to infinity for k outside the ball of radius ϵ about the origin. Then since

$$F(t) = \frac{1}{2}\int e^{2it(k^2+m^2)^{1/2}}(f(k) - f(\Lambda_t^{-1}(k))|\frac{\partial\Lambda_t^{-1}}{\partial k}|)dk$$

and each $f \circ \Lambda_t^{-1}$ has support outside the ball of radius ϵ, and that $\{f \circ \Lambda_t^{-1} | t \in (1,\infty)\}$ is uniformly integrable, we may use the Dunford-Pettis theorem to take the limit under the integral sign. We have that $\lim_{t\to\infty} F(t) = 0$. Now consider the case where f is an arbitrary L^1 function. The argument above applies to $f(k)\chi_{\{|k|\geq\epsilon\}}(k)$ for arbitrary positive ϵ. So we may choose a positive monotonic function $\epsilon(t)$ with $\lim_{t\to\infty}\epsilon(t) = 0$ so that

$$\lim_{t\to\infty}\int e^{2it(k^2+m^2)^{1/2}} f(k)\chi_{\{|k|\geq\epsilon(t)\}}(k)dk = 0 \quad .$$

But clearly the same is true if we reverse the inequality sign in the characteristric function, and so combining the results, we have that $F(t)$ tends to zero in the general case. Applying this to the particular L^1 function considered above, we have the theorem.

In the usual way, if we assume $|\phi_+|^2$ has moments of all orders, the resulting smoothness of $\hat{\phi}_+$ may be used to show that the convergence is faster than any inverse power of t. Also the same holds in any of the smaller spaces in the scale associated to H.

So for large t, $H\phi(\cdot,t)$ and $\dot{\phi}(\cdot,t)$ become more and more orthogonal with more and more nearly equal lengths. Taking them as a basis in the plane they span in $\mathcal{Q}$, we have the following picture for the drift of our diffusion at large time t:

There is a restoring force drift $-H\xi$ at ξ from the ground state process itself. This pushes straight in while the osmotic velocity u pushes straight out. They cancel out on the ring where $u(\xi,t) = H\xi$, and together, they push the field configuration toward this (moving) ring. Meanwhile, the current velocity v pushes the configuration around the ring with a frequency depending on the energy of ϕ. And of course all the while, the plane we are talking about is tumbling about in $\mathcal{Q}$.

A number of questions arise.

First, can one actually solve the stochastic differential equations we are discussing here, and in what sense? Granted a positive answer, we ask more detailed questions.

Second, the tail field of the ground state process is known to be trivial. Of course this means that the tail fields of the coherent state processes are trivial. The simple drift of the general single particle diffusion seems to have such a simple structure that one might expect it too to have a trivial tail field. I don't know how to prove this. Because there is no nice potential theory in infinite dimensions (no Harnack inequality), the question is hard.

Third, is there a reasonable procedure for obtaining the stochastic mechanical diffusions of a free non-relativistic particle in $\mathbf{R}^{d-1}$ as a as some sort of non-relativistic limit of the single particle diffusions discussed here? For instance, ξ_t is a random element of $\mathcal{Q}$, roughly speaking, and one can associate a random point in $\mathbf{R}^{d-1}$ to it by taking $x_t(\omega)$ to be the point where $\xi_t(\omega)$ is most bunched up in some sense—the place where the lump is localized. What sort of processes arise this way?

Fourth, once one has answered the second and third questions, one can ask what happens in the non-relativistic limit if the tail field of the generic single particle diffusion is trivial. In the non-relativistic case, the tail field is not trivial, and it is generated by the *final momentum* of the diffusing particle [7].

Here we will only remark on the first question. The answer is positive; all the single particle diffusions corresponding to solutions ϕ of the Klein-Gordon equation such that $H\phi(\cdot,0)$ and $\dot{\phi}(\cdot,0)$ are in $\mathcal{Q}$ exist as nice diffusions with continuous sample paths in any Banach space $\mathbf{B}$ where the ground state process has continuous sample paths.

This result is easily proved by combining results of Carmona [3] on the ground state process with the technique in [8] for constructing diffusions with singular drifts. The point is that this latter technique required no smoothness of the coefficients; the *finite action condition* imposed there, together with the weak continuity equation make good sense in infinite dimensions, and do hold in the case at hand. To carry out the adaptation of the construction requires nothing more than a straightforward, but lengthly, finite dimensional approximation argument (so that familiar finite dimensional techniques can be used to produce a sequence of approximating diffusions). We will wait until some of the other questions are answered before writing out the details.

Clearly much remains to be done. In the meantime, it is a pleasure to thank Irving Segal for many helpful discussions on field quantization. I would also like to thank Ian Davies and Aubrey Truman for the opportunity to present this paper at Swansea.

BIBLIOGRAPHY

[1] Nelson, E.: *Quantum Fluctuations*, Princeton University Press, Princeton NJ, 1985

[2] Nelson, E.: Field Theory and the Future of Stochastic Mechanics, to appear in the proceedings of the Ascona Conference on Statistical Processes in Classical and Quantum Systems, ed. by S. Albeverio, 1986

[3] Carmona, R.: Measurable Norms and Banach Space Valued Gaussian Processes, Duke Math. Journal, vol. 44, 1977, pp.109-127

[4] Albeverio, S., Høegh-Krohn, R.: Hunt Processes and Analytic Potential Theory on Rigged Hilbert Spaces, Ann. Inst. H. Poincare, vol. 13, 1977, pp.269-291

[5] Segal, I.: Mathematical Problems in Relativistic Physics, A.M.S., Providence RI, 1963

[6] Segal, I: Algebraic Integration Theory, Bull. A.M.S., vol. 71, 1965, pp. 419-489

[7] Carlen, E.: Quantum Scattering and Stochastic Mechanics, to appear in the proceedings of the Ascona Conference on Statistical Processes in Classical and Quantum Physics, ed. by S. Albeverio, 1986

[8] Carlen, E.: Existence and Sample Path Properties of the Diffusions in Nelson's Stochastic Mechanics in *Stochastic Processes in Mathematics and Physics*, ed. by S. Albeverio et. al., Lecture Notes in Mathematics 1158, Springer, Berlin, 1985, pp.25-51

[9] Guerra, F., Ruggerio, P.: A new Interpretation of the Euclidean Markov Field in the Framework of Physical Minkowski Spacetiome, Phys. Rev. Lett., 31, 1973, pp.1022-1025

[10] Segal, I.: Symplectic Structures and the Quantization Problem for Wave Equations, Symposia Mathematica, 14, Bologna, 1974

FLOWS OF NEWTONIAN DIFFUSIONS

M.J. Chappell and K.D. Elworthy
Mathematics Institute, University of Warwick, Coventry CV4 7AL

§0. Introduction

A. Consider a particle, mass m, with state space a Riemannian manifold M (which we shall usually take to be an open domain of $\mathbb{R}^n$) moving under a force $-\nabla V$ and subjected to random fluctuations. With assumptions on the fluctuations, and under certain regularity conditions, Nelson showed, [14], that a generalization of Newton's equations leads to the motion of the particle being markovian with generator A for

$$A = \tfrac{1}{2m}\Delta + b_t \tag{1}$$

where Δ is the Laplace-Beltrami operator of M and the drift $\{b_t : t \in \mathbb{R}\}$ is a time dependent vector field (possibly with singularities) obtainable from V, the Riemannian metric, and initial conditions by means of some non-linear equations. Later this property, and the equations, were derived by Yasue, and by Guerra and Morato, using stochastic versions of classical mechanical variational principles, see [14].

The resulting Markov process has been studied with this sort of interpretation in connection with a variety of phenomena: physical, biological, and sociological, e.g. see [1], and articles by Carlen, by Truman, and by Yasue in this volume. However most of these studies are concerned with quantum mechanics. Indeed the Schrödinger equation with potential V arises as a way of linearising the non-linear equations for b and it is natural to attempt to interpret properties of the Markov process in terms of those of solutions of this Schrödinger equation and conversely e.g. see [4], [13], [18].

The correspondence with the Schrödinger equation

$$i\hbar \frac{\partial}{\partial t}\psi_t(x) = \left(-\frac{\hbar^2}{2m}\Delta + V(x)\right)\psi_t(x) \tag{2}$$

is as follows:

Let ψ satisfy (2). Define the time dependent vector fields u, (the <u>osmotic velocity</u>), and v (the <u>current velocity</u>) by

$$u_t(x) = \begin{cases} \frac{\hbar}{m}\,\mathrm{Re}\left(\frac{\nabla\psi_t(x)}{\psi_t(x)}\right) & \psi_t(x) \neq 0 \\ 0 & \text{otherwise} \end{cases}$$

$$\text{and}\quad v_t(x) = \begin{cases} \frac{\hbar}{m}\ \mathrm{Im}\left(\frac{\nabla\psi_t(x)}{\psi_t(x)}\right) & \psi_t(x) \neq 0 \\ 0 & \text{otherwise} \end{cases} .$$

Set

$$b_t(x) = u_t(x) + v_t(x) . \tag{3}$$

Under suitable regularity conditions there exists a Markov process with generator $\frac{\hbar}{2m}\Delta + b_t$ with distribution ρ_t at time t given by

$$\rho_t(x) = |\psi_t(x)|^2$$

and whose trajectories almost surely do not hit the zeros of ψ_t at time t (the "nodal surfaces"). See [3], [20]. Such regularity will be assumed throughout.

B. Such a Markov process can be represented as the solution of a stochastic differential equation

$$dx_t = b_t(x_t)dt + \frac{\hbar}{m} A(x_t)dt + \sqrt{\frac{\hbar}{m}}\, X(x_t)\circ dB_t \tag{4}$$

provided the solutions to the Stratonovich equation

$$dy_t = A(y_t)dt + X(y_t)\circ dB_t \tag{5}$$

form a Brownian motion on M. Here A is a vector field on M and for each given $y \in M$

$$X(y) : \mathbb{R}^p \to T_yM' \tag{6}$$

maps $\mathbb{R}^p$, for some p, linearly into T_yM, thereby injecting the increments of the p-dimensional Brownian motion $\{B_t : t \geq 0\}$ into the tangent spaces to the manifold.

There are many choices of A, p, and X which will give Brownian motion. For example when $M = \mathbb{R}$ a natural choice is $p = 1$, $A \equiv 0$, and $X(y)$ the identity, giving $dy_t = dB_t$. However the periodic choice $p = 2$, $A \equiv 0$, and $X(y) = (\cos y, -\sin y)$, giving

$$dy_t = (-\sin y_t)dB_t^1 + (\cos y_t)dB_t^2 \tag{7}$$

is equally suitable. The choice of X, A gives additional structure to the situation, and physically means making some assumptions on how the fluctuations are being introduced into the system: i.e. on the 'background fields'. As observed by Stroock the coefficients A and X can be obtained from the joint distribution of $\{(x_t, B_t) : t \geq 0\}$ on $M \times \mathbb{R}^p$, by considering it as a Markov process and looking at its generator using Itô's formula, for example. The joint distributions of (x_t, y_t) on $M \times M$ would do equally well, as would the '2 point motion' on $M \times M$ obtained by solving (5) at 2 different points, as used by Baxendale [2].

C. If the coefficients were smooth and M were compact the equation (4) would have a flow

$$F_t(-\ ,\omega) : M \to M \qquad\qquad t \geq 0$$

of diffeomorphisms, for $\omega \in \Omega$: the probability space associated to $\{B_t : t \geq 0\}$. See [8], [11], [12]. For x_0 in M the process $\{F_t(x_0,-) : t \geq 0\}$ is then the solution of (4) starting from x_0 . (The same is true for x_0 an M-valued random variable independent of $\{B_t : t \geq 0\}$) . In the smooth case for M not necessarily compact local flows exist by a result of Kunita, see [12], [8], [11], and we can apply this to our situation by considering the equation for (y_t,t) on space-time $M \times [0,\infty)$ with the nodal sets removed. In these circumstances open sets may be sent by $F_t(-,\omega)$ to infinity (or into nodal sets) in finite time even though this almost surely never happens to individual points: the system is complete, but may not be <u>strongly complete</u> (≡ strictly conservative). We show in §2 that this does indeed happen even for the ground state of the hydrogen atom with the most obvious choice of X , A .

An important property of flows is the behaviour of the distances $d(F_t(x_0,\omega), F_t(y_0,\omega))$ between two solutions for large time. This infinitesimally is measured by Lyapunov exponents as in Ruelle's ergodic theory for ordinary dynamical systems. Some remarks about these for ground state flows of stochastic mechanical systems were made in [7] and these are further developed here. Even when there is not stochastic completeness, derivatives in probability of $F_t(x_0,\omega)$ with respect to $x_0 \in M$ exist, [11], to give a process $v_t(\omega) = T_{x_0}F_t(-,\omega)(v_0)$ for each tangent vector v_0 at x_0 . There are then the exponents

$$\bar{\lambda}(v_0,\omega) = \overline{\lim_{t\to\infty}} \frac{1}{t} \log|v_t(\omega)|$$

and $$\underline{\lambda}(v_0,\omega) = \underline{\lim_{t\to\infty}} \frac{1}{t} \log|v_t(\omega)| \quad . \tag{8}$$

For equations like (4) with smooth coefficients on compact M, Carverhill [5] showed that for almost all points (x,ω) of $M \times \Omega$ there is a filtration of T_xM by linear subspaces

$$0 = V^{r+1}_{(x,\omega)} \subset \ldots \subset V^2_{(x,\omega)} \subset V^1_{(x,\omega)} = T_xM$$

with real numbers $\lambda^r < \ldots < \lambda^1$ such that for $j = 1$ to r

$$v \in V^j_{(x,\omega)} - V^{j+1}_{(x,\omega)} \Rightarrow \bar{\lambda}(v,\omega) = \underline{\lambda}(v,\omega) = \lambda^j \quad .$$

Moreover if $\lambda_\Sigma = \sum_j \lambda^j \dim(V^j_{(x,\omega)} / V^{j+1}_{(x,\omega)})$

then λ_Σ is independent of (x,ω) and

$$\lambda_{\Sigma} = \lim_{t\to\infty} \frac{1}{t} \log \det T_x F_t(-,\omega) \ . \tag{9}$$

This comes from the multiplicative ergodic theorem, following Ruelle, and continuing with a stochastic version of Ruelle's work Carverhill gave the stable manifold theorem: if $\lambda^j < 0$ then for almost all $(x,\omega) \in M \times \Omega$ the stable manifold

$$V^j_{(x,\omega)} = \{y \in M : \overline{\lim_{t\to\infty}} \frac{1}{t} \log d(F_t(\omega,x), F_t(\omega,y)) \le \lambda^j\} \tag{10}$$

is an immersed manifold in M tangent to $V^j_{(x,\omega)}$ at x.

Here we will show how some of these results go over to stochastic mechanical flows associated to superpositions of bound states of (2). We also briefly contrast them with what can happen for a simple class of scattering states. Finally we consider the ground state flow for the hydrogen atom. This has some special features, for example a flow does not exist on the obvious state space: however we give Chappell's result that the top exponent is negative. This goes a long way towards explaining the near coalescence observed in the computer simulation, Figure 1, of the article by Durran and Truman in this volume.

The picture that seems to emerge is that, for a process obtained from the evolution ψ_t of a superposition of bound states, if an observer looks at time t at the process it has density $|\psi_t(-)|^2$. However if he then follows the progress of this picture as it moves in the future under the influence of a generic sample path of fluctuations he will notice it concentrate rapidly so that all its mass is around one or more moving points. The exponential rate of concentration (at least measure theoretically) is controlled by the osmotic velocity, and in the case of ground state processes is proportional to the quantum mechanical kinetic energy (at least for the most natural choices of X, A).

§1. Ergodic theory for superpositions of bound states

A. Let ψ^j for $j = 1$ to k be eigenfunctions of our Schrödinger operator with eigenvalues E_j, $j = 1$ to k:

$$\left(-\frac{\hbar^2}{2m}\Delta + V\right)\psi^j = E_j \, \psi^j \ . \tag{11}$$

Set $\psi_t(x) = \sum_{j=1}^{k} \alpha_j \, e^{iE_j t} \psi^j(x) \qquad t \in \mathbb{R}$

where $(\alpha_1,\ldots,\alpha_k) \in \mathbb{C}^k$ is fixed. Taking $M = \mathbb{R}^n$, represent the stochastic mechanical diffusion corresponding to $\{\psi_t : t \ge 0\}$ by a solution of the equation

$$dx_t = b_t(x_t)dt + \sqrt{\frac{\hbar}{m}} \, . dB_t \tag{12}$$

where $\{B_t : t \ge 0\}$ is a Brownian motion on M, and b is given by (3). We assume that V is also sufficiently regular so that each ψ^j is smooth and of finite energy (i.e. $\nabla\psi^j \in L^2$) . Then (12) coupled with the equation $d\tau_t = dt$ has strong solutions $\{(x_t,\tau_t) : t \geq 0\}$ in $\mathbb{R}^n \times [0,\infty) - N$ where $N = \{(x,t):\psi_t(x) = 0\}$, and admits at least a local flow $(x,\tau) \to (F_t(x,\omega) , t + \tau)$. There is a derivative $T_xF_t(-,\omega) : \mathbb{R}^n \to \mathbb{R}^n$ in the L^0 sense [11] which is given by $v_0 \to v_t(\omega)$ with

$$dv_t = Db_t(x_t)(v_t)dt . \tag{13}$$

We will show that Ruelle's method of dealing with quasi-periodic time dependent dynamical systems [15], implies the existence of a Lyapunov spectrum for $\{T_xF_t(-,\omega) : t \geq 0\}$ and in particular the existence, almost surely, of

$$\lim_{t\to\infty} \frac{1}{t} \log|v_t| .$$

B. Let T^k denote the k-torus. Define

$$g^t : T^k \to T^k \qquad\qquad t \in \mathbb{R}$$

by $g^t(q_1,\ldots,q_k) = (e^{itE_1}q_1,\ldots,e^{itE_k}q_k)$ considering each $q_j \in S^1$ as a complex number. Let S be the closure of the orbit of 1 under $\{g^t : t \in \mathbb{R}\}$. This is another torus; let σ be its normalized Haar measure, so the restriction of g^t to S is measure preserving, and ergodic, and even minimal (e.g. see [19]).

For $q \in S$ let $\{\psi(q)_t : t \in \mathbb{R}\}$ be the evolution of the wave function $\psi(q)_0$ where

$$\psi(q)_0(x) = \sum_{j=1}^{k} \alpha_j q_j \psi^j(x) .$$

Then $\psi(q)_t(x) = \psi(g^t q)_0$

and the corresponding stochastic mechanical process is given by

$$dx_t = b(q)_t(x)dt + \sqrt{\frac{\hbar}{m}}\, dB_t \tag{14}$$

for $b(q)_t$ defined as in (3).

Then $b(q)_t(x) = b(g^t q)_0(x)$.

Let $N(q)$ be the nodal set of $\psi(q)_0$ and let Γ be a connected component of $(R^n \times S)' \equiv \mathbb{R}^n \times S - \{(x,q) : x \in N(q)\}$. One way to proceed would be to consider the (degenerate) stochastic dynamical system on Γ given by (14) rewritten as

$$dx_t = b(q_t)_0(x_t)dt + dB_t \tag{14$'$}$$

and

$$dq_t = (iE_1,\ldots,iE_k).q_t dt \tag{15}$$

using the multiplication of $\mathbb{C}^k$ and inclusion of S in $\mathbb{C}^k$. We might then hope to obtain the results of [5] for this system i.e. Lyapunov spectrum and stable manifold theorem. A slightly different way and one closer to that followed by Ruelle is to replace the basic probability space $(\Omega, F, \mathbb{P})$ by $(\tilde{\Omega}, \tilde{F}, \tilde{\mathbb{P}})$ where $\tilde{\Omega} = \Omega \times S$, $\tilde{F} = F * \text{Borel } S$, and $\tilde{\mathbb{P}} = \mathbb{P} \otimes \sigma$. One then considers (14), or rather its space-time version with N removed, as an S.D.E. with this additional element of randomness. In either case there is a measurable

$$\Phi_t : \Gamma \times \Omega \to \Gamma \times \Omega$$

of the form $\Phi_t(x,q,\omega) = (\tilde{F}_t(x,q,\omega), g^t q, \theta_t \omega)$ where $\theta_t : \Omega \to \Omega$ is the shift (taking Ω as classical Wiener space) and $\tilde{F}$ is a measurable solution flow for (14)', (15) off the nodal sets, see [11] : a smooth version may perhaps not exist.

Since the Markov process determined by (14)', (15) has invariant measure $|\psi(q)_0(x)|^2 dx \otimes \sigma(dq)$ the semi-group of transformations Φ_t preserves the measure μ given by

$$|\psi(q)_0(x)|^2 dx \otimes \sigma(dq) \otimes \mathbb{P}(d\omega) .$$

In fact μ is ergodic: as can be seen by observing that for any Dirac measure $\delta(x_0, q_0)$ on Γ the measure $\delta(x_0, q_0) \otimes \mathbb{P}$ gets mapped by Φ_t to one whose projection onto M has support of the whole of $\Gamma \cap (\mathbb{R}^n - N(g^t q_0))$.

C. Let $T_x \tilde{F}_t(-,q,\omega)(v_0)$ denote the partial derivative in measure of F_t in the direction $v_0 \in \mathbb{R}^n$. To apply the multiplicative ergodic theorem as in [5] we need the following key lemma:

<u>Lemma</u> 1C For each $t > 0$ both quantities

$$\log^+ ||T_x \tilde{F}_t(-,q,\omega)|| \quad \text{and} \quad \log^+ ||(T_x \tilde{F}_t(-,q,\omega))^{-1}||$$

are integrable with respect to μ .

<u>Proof</u> (Chappell [9]). We can take the norms to be operator norms on $L(\mathbb{R}^n;\mathbb{R}^n)$. Define $V_t \equiv V_t(q,\omega)$ by

$$V_t = T_{x_0} F_t(-,q,\omega)(v_0)$$

and

$$\beta_s \equiv \beta(q)_s : \mathbb{R}^n \to \mathbb{R}^n$$

by $\beta_s = D(b(q_s)_0)(x_s)$.

Then $dV_t = D(b(q_t)_0)(x_t)(V_t)dt$ for $q_t = g^t q$. By Itô's formula, if $||v_0|| = 1$,

$$\log ||V_t||^2 = \int_0^t 2\langle V_s, \beta_s V_s \rangle \, ||V_s||^{-2} ds .$$

Consequently $\log||T_x F_t(-,q,\omega)|| \le \int_0^t \sup_{\{v_0:||v_0||=1\}} |\langle V_s,\beta_s V_s\rangle| |V_s|^{-2} ds$.

This shows, via the Cauchy-Schwarz inequality that

$$\int \log^+ ||T_x F_t(-,q,\omega)|| \mu(dx,dq,d\omega) \le$$

$$\le \int\int_0^t ||\beta_s(q)(\tilde{F}_s(x,q,\omega))|| ds \ \mu(dx,dq,d\omega)$$

$$= \int\int_0^t ||\beta(q)_0(x)|| ds \ \mu(dx,dq,d\omega)$$

(by Φ_t invariance)

$$= t\int_{(R^n\times S)'} ||D(b(q)_0)(x)|| \ |\psi(q)_0(x)|^2 dx \ \sigma(dq) .$$

For the first part it therefore suffices to show that each element of the matrix of the linear transformation $|\psi(q)_0(x)|^2 \ D(b(q)_0)(x)$ lies in $L^1 \ ((\mathbb{R}^n\times S)' ,$ $dx \otimes \sigma(dq))$. These entries are the sums of the real and imaginary parts

$$(I + R)(\bar{\psi} \frac{\partial^2}{\partial x_i \partial x_j} \psi - \frac{\bar{\psi}}{\psi} \frac{\partial\psi}{\partial x_i} \frac{\partial\psi}{\partial x_j})$$

where $\psi = \psi(q)_0$. Now the finite energy condition implies that $|\frac{\bar{\psi}}{\psi} \frac{\partial\psi}{\partial x_i} \frac{\partial\psi}{\partial x_j}|$ is integrable over $(\mathbb{R}^n \times S)'$, and also that $|\bar{\psi} \frac{\partial^2\psi}{\partial x_i \partial x_j}|$ is since

$$||\bar{\psi} \frac{\partial^2\psi}{\partial x_i\partial x_j}||_0 = ||\sum_{\ell,m=1}^{k} c_{\ell,m}(q)\psi^\ell \ \frac{\partial^2\psi^m}{\partial x_i\partial x_j}||_0$$

$$\le \sum_{\ell,m=1}^{k} |c_{\ell,m}(q)| ||\psi^\ell||_1 \ ||\frac{\partial^2\psi^m}{\partial x_i\partial x_j}||_{-1}$$

where the $c_{\ell,m}$ are bounded functions of q and the norms are those of the Sobolev spaces $H^0(\mathbb{R}^n)$, $H^1(\mathbb{R}^n)$ and $H^{-1}(\mathbb{R}^n)$ respectively. This proves the integrability of the first quantity.

For the second quantity observe that if $A_t = (T_x\tilde{F}_t(-,q,\omega))^{-1}$ then

$$dA_t = -A_t \ D(b(q_t)_0)(x_t)(-)dt .$$

By the equivalence for operators on $\mathbb{R}^n$ between the operator norm $||\quad||$, and the Hilbert-Schmidt norm $||\quad||_2$, with inner product $\langle\langle\quad\rangle\rangle$ we have
$\lim_{t\to\infty} \frac{1}{t} \log||A_t|| = \lim_{t\to\infty} \frac{1}{t} \log||A_t||_2$
and $\log||A_t||_2 = -\int_0^t \langle\langle A_s,A_s D(b(q_s)_0)(x_s)(-)\rangle\rangle \ ||A_s||_2^{-2} ds$

giving $\log^+||A_t||_2 \le \text{const.} \int_0^t ||Db(q_s)_0(x_s)||ds$ from which the result follows as before. //

From this lemma we can apply the multiplicative ergodic theorem as did Carverhill [5], to obtain a Lyapunov spectrum for $T_x\tilde{F}_t(-,q,\omega)$ as described in the introduction, defined for almost all (x,q,ω) in $\Gamma \times \Omega$. From this there is a set Q of full measure in S, invariant under g^t such that the spectrum is defined with $q \in Q$ for almost all $(x,\omega) \in (\mathbb{R}^n-N(1)) \times \Omega$.

From the minimality of g^t on S we see that we can take $Q = S$: use the lower semicontinuity in q of $\overline{\lim}\, \frac{1}{t} \log|T_x\tilde{F}_t(-,q,\omega)v_0|$ for each (x,ω) and the upper semicontinuity of the corresponding lower limit together with the fact that Q is dense in S. Thus we can take $q = 1$ and get a Lyapunov spectrum for $T_xF_t(-,\omega)$ defined at almost all (x,ω) in $(\mathbb{R}^n-N) \times \Omega$.

D. When the system is strongly complete, so the flow exists as a smooth map, λ_Σ shows how the measure of a subset grows, or decreases, as it is moved by the flow. When the system is complete but not strongly complete this still holds:

<u>Proposition</u> 1D Let $dx_t = X(x_t)\circ dB_t + A(x_t)dt$ (16)

be a stochastic dynamical system with smooth coefficients on a smooth manifold M. Suppose that it is complete. Let F_t, $t \ge 0$ be the partially defined smooth partial flow ([12], [8], [11]) with explosion time map

$$\tau : M \times \Omega \to (0,\infty] .$$

Then τ is $\mu \otimes \mathbb{P}$ almost-surely finite for any locally finite Borel measure μ on M. Consequently there is a measurable solution flow

$$\tilde{F}_t : M \times \Omega \to M \qquad t \ge 0$$

such that $\tilde{F}_t(-,\omega) : M \to M$ is C^∞ on an open subset $M(t)(\omega)$ of M which has full measure for all such μ.

<u>Proof</u> It suffices to consider probability measures μ on M with compact support. For such take $u : \Omega \to M$ independent of the driving Brownian motion with μ as distribution (if necessary by enlarging Ω). Let $\{x_t : t \ge 0\}$ be the solution to (16) with initial distribution $x_0 = u$. For any smooth $\phi: M \to \mathbb{R}$ let x_t^ϕ and F_t^ϕ correspond to x_t and F_t but for (16) replaced by the corresponding equation with coefficients multiplied by ϕ.

For any $\varepsilon > 0$ and $T > 0$ there is a compact set K such that $x_t(\omega) \in K$ for $0 \le t \le T$ with probability greater than $1 - \varepsilon$. Choose ϕ with compact support and identically one on K. Then $x_t^\phi = x_t$ for $0 \le t \le T$ with probability greater than $1 - \varepsilon$. However, for $t \ge 0$,

$$x_t^\phi(\omega) = F_t^\phi(u(\omega),\omega)$$

for example as in [11] Theorem IX 3B.

Setting $M(t)(\omega) = \{x \in M : \tau(x,\omega) > t\}$ it follows that there is probability greater than $1-\varepsilon$ that $u(\omega)$ lies in $M(T)(\omega)$ with $x_t(\omega) = F_t(u(\omega))(\omega)$ for $0 \le t \le T$. Since ε was arbitrary $M(T)(\omega)$ has full μ-measure, as required for almost all $\omega \in \Omega$.

The version $\tilde{F}_t$ can be defined to be equal to F_t on each $M_t(\omega)$ and otherwise to be given by $\tilde{F}_t(x,\omega) = x$. //

<u>Corollary</u> 1D. For a complete stochastic dynamical system as in the proposition there is a measurable solution flow $\tilde{F}_t : M \times \Omega \to M$ which almost surely leaves the Lebesgue measure of M quasi-invariant. For a Riemannian metric on M the change in Riemannian volume is given by the Jacobian determinant of the derivative in probability of $\tilde{F}_t(-,\omega)$. In particular when a Lyapunov spectrum exists λ_Σ represents the almost sure exponential rate of change of volume under the flow. //

E. Next we compute $\lambda_\Sigma(x_0)$. Since

$$\frac{d}{dt} \det T_{x_0}\tilde{F}_t(-,\omega) = \operatorname{div}(b_t)x_t(\omega)$$

for $x_t(\omega) = F_t(x_0,\omega)$,

$$\lambda_\Sigma(x_0) = \lim_{t\to\infty} \frac{1}{t}\int_0^t \operatorname{div}(b_s)(x_s)ds \qquad \text{a.s.}$$

$$= \frac{1}{|\Gamma|}\int_\Gamma \operatorname{div}(b(q)_0)(x)|\psi(q)_0(x)|^2\, dx\sigma(dq) \tag{17}$$

where Γ is the component of $(x_0,1)$ in $(\mathbb{R}^n \times S)'$ and

$$|\Gamma| = \int_\Gamma |\psi(q)_0(x)|^2\, dx\, \sigma(dq) .$$

The following lemma has some independent interest:

<u>Lemma</u> 1E
$$\lim_{t\to\infty} \frac{1}{t}\int_0^t \operatorname{div} v_s(x_s(\omega))ds$$
$$= \lim_{t\to\infty} \frac{1}{t}\int_0^t \langle u_s(x_s(\omega)),v_s(x_s(\omega))\rangle ds$$
$$= 0$$

for almost all (x_0,ω).

<u>Proof</u> (As suggested by J.C. Zambrini) By the ergodic theorem

$$\frac{1}{t}\int_0^t \operatorname{div} v_s(x_s(\omega))ds \to \frac{1}{|\Gamma|}\int_\Gamma \operatorname{div}(v(q)_0)(x)|\psi(q)_0(x)|^2 dx\, \sigma(dq) .$$

For $\rho(q)_t(x) = |\psi(q)_t(x)|^2$ there is the continuity equation

$$\frac{\partial\rho}{\partial t} + \mathrm{div}(\rho v) = 0 . \tag{18}$$

Now $\int_\Gamma \frac{\partial}{\partial t} \rho(q)_t(x)\, dx\sigma(dq)$

$$= \frac{\partial}{\partial t}\int_\Gamma \rho(g^t q)_0(x)dx\ \sigma(dq)$$

$$= 0$$

by the invariance of σ and Γ under g^t . Thus

$$\frac{1}{t}\int_0^t \mathrm{div}\ v_s(x_s(\omega))ds \to -\frac{1}{|\Gamma|}\int_\Gamma <v(q)_0(x),\nabla(\rho(q)_0)(x)>\ dx\ \sigma(dq) . \tag{19}$$

Also $\frac{1}{|\Gamma|}\int_\Gamma <v(q)_0(x),\nabla\rho(q)_0(x)>dx\ \sigma(dq)$

$$= \frac{2}{|\Gamma|}\int_\Gamma <v(q)_0(x),u(q)_0(x)>\rho(q)_0(x)dx\ \sigma(dq) \tag{20}$$

$$= \lim_{t\to\infty}\frac{2}{t}\int_0^t <v_s(x_s(\omega)),u_s(x_s(\omega))>ds \quad .$$

Now, for all $t \geq 0$, $\frac{1}{|\Gamma|}\int_\Gamma <v(q)_0(x),u(q)_0(x)>\rho(q)_0(x)dx\ \sigma(dq)$

$$= \frac{1}{t}\int_0^t \mathbb{E}\ \frac{1}{|\Gamma|}\int_\Gamma <v(q)_0(x),u(q)_0(x)>\rho(q)_0(x)dx\ \sigma(dq)$$

$$= \frac{1}{t|\Gamma|}\int_\Gamma \mathbb{E}\int_0^t <v(q)_s(\tilde{F}_s(x,q)),u(q)_s(\tilde{F}_s(x,q))>\rho(q)_0(x)dx\ \sigma(dq)ds$$

$= I_t$, say.

Also since $v = \frac{1}{2}(Dx + D_*x)$, the average of the mean forward and mean backward velocities, it follows from the definitions of these quantities and of the Stratonovich integral (e.g. $\circ\, dx = \frac{1}{2}(dx + d_*x)$ as in [14]) that

$$\mathbb{E}\int_0^t <u_s(x_s),v_s(x_s)>ds$$

$$= \mathbb{E}\int_0^t u_s(x_s)\circ dx_s$$

$$= \tfrac{1}{2}\,\mathbb{E}\ \{\log|\psi_t(x_t)|^2 - \log|\psi_0(x_0)|^2$$

$$- \int_0^t \frac{\partial}{\partial r}\log|\psi_r(x_s)|^2\Big|_{r=s} ds\}$$

since $u = \frac{1}{2}\nabla\log|\psi|^2$. Thus, by the invariance of μ ,

$$|\Gamma|\ I_t = -\frac{1}{2t}\int_\Gamma\int_0^t \frac{\partial}{\partial r}|\psi(g^{-s}q)_r|^2\Big|_{r=s} ds\ \sigma(dq)dx$$

$$= -\frac{1}{2t}\int_\Gamma\int_0^t \frac{\partial}{\partial s}|\psi(q)_s|^2 ds\sigma(dq)dx$$

by invariance of σ under g^s . Thus

$$I_t = -\frac{1}{2t|\Gamma|}\int_\Gamma (|\psi(q)_t|^2 - |\psi(q)_0|^2)\sigma(dq)dx$$

$$= 0 .$$

The result follows from (19) and (21). //

For the following we need in particular the weak conditions on the potential V that $V_+ \in K_n^{loc}$ and $V_- \in K_n$, for K_n, as in [17] for example.

Proposition 1E. For almost all $(x_0,\omega) \in (\mathbb{R}^n - N(1)) \times \Omega$

$$\lambda_\Sigma(x_0) = \lim_{t\to\infty} \frac{1}{t}\int_0^t \det\ T_{x_0}F_s(-,\omega)ds$$

$$= -\frac{2}{|\Gamma|}\int_\Gamma |u(q)_0(x)|^2\ |\psi(q)_0(x)|^2 dx\ \sigma(dq) \tag{22}$$

where Γ is the component of $(x_0,1)$ in $\mathbb{R}^n \times S - \{(x,q):x \in N(q)\}$.

Proof From (9) and the lemma, with probability 1,

$$\lambda_\Sigma(x_0) = \frac{1}{|\Gamma|}\int_\Gamma |\psi(q)_0(x)|^2 \operatorname{div}\ u(q)_0(x)\ dx\ \sigma(dq) .$$

Now for fixed $q \in S$ the vector field $|\psi(q)_0(x)|^2 u(q)_0(x)$ on $\mathbb{R}^n$-$N(q)$ is just the restriction of the gradient of the function $x \to |\psi(q)_0(x)|^2$ which has $N(q)$ as the set on which it attains its absolute minimum. The vector field is therefore complete on $\mathbb{R}^n$-$N(q)$ provided it was complete on $\mathbb{R}^n$. However our assumptions on V imply that $|\psi(q)_0(x)|^2 \to 0$ as $x \to \infty$ for each q ([17] Theorem C.3.1.), and so this completeness is true. Also $\operatorname{div}(|\psi(q)_0|^2 u(q)_0)$ is Lebesgue integrable over $\mathbb{R}^n$-$N(q)$, by arguing as in the proof of lemma 1C. Thus we can apply the divergence theorem for each q to see

$$\lambda_\Sigma(x_0) = -\frac{1}{|\Gamma|}\int_\Gamma \langle u(q)_0(x),\nabla|\psi(q)_0|^2(x)\rangle dx\sigma(dq)$$

giving the required result since $u(q)_0 = \frac{1}{2}\nabla\log|\psi(q)_0|^2$. //

Remark Since $u = \nabla\log|\psi|$ the expression (22) can be written

$$\lambda_\Sigma(x_0) = -\frac{2}{|\Gamma|}\int_\Gamma |\nabla|\psi(q)_0|(x)|^2 dx\sigma(dq)$$

which shows how it reduces a constant times the kinetic energy when ψ is the ground state, as in [7].

F. Observe that for scattering states the results of Shucker [16] and of Carlen [4] suggest that (for state space $\mathbb{R}^n$) the situation is quite different. Indeed under their conditions $\frac{1}{t}x_t$ converges almost surely as $t \to \infty$ so one might expect solutions from two different points to grow apart linearly (or end up moving essentially parallel to each other). The free evolutions of a complex Gaussian wave function are described in §16 of [14]. The stochastic differential equations are linear and in the case of the "one-slit" process

$$u_t = -\frac{2\lambda^2}{4\lambda^4+t^2} x_t$$

$$v_t = \frac{t}{4\lambda^4+t^2} x_t$$

so that $d(F_t(x_0,\omega), F_t(y_0,\omega)) \to \infty$ linearly as $t \to \infty$. There is no obvious ergodic theory to give the existence of a Lyapunov spectrum, but certainly in this case

$$\lim_{t\to\infty} \frac{1}{t} \log \det T_x F_t(-,\omega) = 0 \qquad \text{a.s.}$$

§2. The ground state process for the hydrogen atom

A. For the hydrogen atom $V(x) = -Ze^2/|x|$ on $\mathbb{R}^3-\{0\}$. The ground state is given

$$\psi_t(x) = N \exp(-iEt/\hbar)\exp(-|x|/a)$$

where $a = \hbar^2/(me^2Z)$, for Ze the charge of the nucleus, and $E = -\hbar^2/(2ma^2)$, with N a normalizing constant. The corresponding process [18], [13] is given by

$$dx_t = -\frac{\hbar}{ma}\frac{x_t}{|x_t|}\, dt + \left(\frac{\hbar}{m}\right)^{\frac{1}{2}} dB_t . \tag{23}$$

As we have been doing, take $\hbar = 1$, $m = 1$.

B. Equation (23) is complete in $\mathbb{R}^3-\{0\}$ i.e. its trajectories from any point of $\mathbb{R}^3-\{0\}$ almost surely never meet the origin: see [13]. However it is not strongly complete:

<u>Proposition</u> There is no smooth flow $F_t(-,\omega) : \mathbb{R}^3-\{0\} \to \mathbb{R}^3-\{0\}$ for (23), defined for almost all $\omega \in \Omega$. However there does exist, for almost all ω,

$$\hat{F}_t(-,\omega) : \mathbb{R}^3 \to \mathbb{R}^3 \qquad t \geq 0$$

which is globally Lipschitz and such that $\{\hat{F}_t(x,-) : t \geq 0\}$ solves (23), with initial point x, for each $x \in \mathbb{R}^3-\{0\}$, where any trajectories which reach 0 are replaced by ones constantly at x.

<u>Proof</u> From (23) if $x_0 \in \mathbb{R}^3-\{0\}$ with $|x_0| = \delta$, say,

$$|x_t-B_t| \leq \delta + t/a \qquad t \geq 0 .$$

Now there is a positive probability that $|B_t| > \delta + t/a$. Thus if a flow existed it must have positive probability of mapping a sphere radius δ about 0 into some disc which does not contain 0: this is topologically impossible for a continuous map.

To obtain the Lipschitz flow on $\mathbb{R}^3$ let $x_0, y_0 \in \mathbb{R}^3-\{0\}$, with x_t, y_t, $t \geq 0$, the corresponding solutions to (23). From (23) it is easy to see that $|x_t - y_t| \le |x_0 - y_0|$ for all t. The smooth partial flow F_t defined on a dense open subset of $\mathbb{R}^3-\{0\}$ by Proposition 1D therefore has a Lipschitz extension $\hat{F}_t$ as required. //

The maps $\hat{F}_t(-,\omega)$ are not expected to be diffeomorphisms, but it seems likely that they are sufficiently smooth to allow the stable manifold theorem to hold: we proceed to show that the top Lyapunov exponent is negative.

Proposition (Chappell [9]) For the ground state process given by (23) $\lambda^1 < 0$.

Proof The derivative process $\{v_t : t \ge 0\}$ satisfies

$$dv_t = -\frac{1}{a}\left(\frac{v_t}{|x_t|} - \langle x_t, v_t\rangle \frac{x_t}{|x_t|^3}\right)dt \tag{24}$$

$$\text{where } \log|v_t| = \log|v_0| + \frac{1}{a}\int_0^t \frac{1}{|x_s|}\left(\frac{\langle x_s, v_s\rangle^2}{|x_s|^2|v_s|^2} - 1\right)ds \tag{24}$$

$$\le \log|v_0|$$

by the Cauchy-Schwarz inequality, so the exponents are certainly non-positive. To prove they are negative we use Carverhill's method [6] of going to the sphere bundle: in this case considering the Markov process

$$\{(x_t, v_t/|v_t|) \ : \ t \ge 0\} \text{ on } (\mathbb{R}^3-\{0\}) \times S^2 .$$

Since S^2 is compact by the Markov-Kakutani fixed point theorem, there exists a probability measure μ on this sphere bundle, invariant for the Markov process, and projecting onto the measure $|\psi_0(x)|^2 dx$ on $\mathbb{R}^3-\{0\}$, see [10]. In fact from Crauel's Corollary 3 (ii) of [10] there exists such μ with, for a dense set of $v_0 \in S^2$,

$$\lambda^1 = \lim_{t\to\infty} \frac{1}{t}\log|v_t|$$

$$= \lim_{t\to\infty} \frac{1}{at}\int_0^t \frac{1}{|x_s|}\left(\left\langle \frac{x_s}{|x_s|}, \frac{v_s}{|v_s|}\right\rangle^2 - 1\right)ds$$

$$= \lim_{t\to\infty} \frac{1}{at}\int_0^t \mathbb{E}\int_{\mathbb{R}^3-\{0\}} \frac{1}{|F_s(x,\omega)|}\left(\left\langle \frac{F_s(x,\omega)}{|F_s(x,\omega)|}, \frac{T_xF_s(-,\omega)v_0}{|T_xF_s(-,\omega)v_0|}\right\rangle^2 - 1\right) |\psi_0(x)|^2 dx\, ds$$

(since λ^1 is almost surely constant in (x_0,ω))

$$= \frac{1}{a}\int_{(\mathbb{R}^3-\{0\})\times S^2} \frac{1}{|x|}\left(\left\langle \frac{x}{|x|}, \alpha\right\rangle^2 - 1\right)\mu(dx,d\alpha) . \tag{25}$$

Here we have applied Crauel's result with $M = S^2$, and taking his probability space to be our $(\mathbb{R}^3-\{0\}) \times \Omega$ with measure $|\psi_0(x)|^2 dx \otimes \mathbb{P}(d\omega)$ and shift given by

$(F_t(x,\omega),\theta_t\omega)$. The measure μ can be obtained as an accumulation point of the orbit of the Dirac measure δ_{v_0} on S^2 under its evolution in $L^\infty((\mathbb{R}^3-\{0\}) \times \Omega; \text{Prob } S^2)$, averaged out with $\mathbb{P}$ to ensure independence from ω .

Formula (25) immediately shows that $\lambda^1 < 0$ unless μ has support in $\{(x,\alpha) : x = \lambda\alpha \text{ some } \lambda \in \mathbb{R}\}$. However this set contains no subset invariant under the sphere bundle flow as is immediate from equation (24).

ACKNOWLEDGEMENTS

This work was helped a lot by our contacts with E. Carlen and J.C. Zambrini, and was partially supported by SERC grants GR/C 46659, GR/C 60860, GR/D 23404.

REFERENCES

1. Albeverio, S., Blanchard, Ph., and Hoegh-Krohn, R. (1984). Newtonian diffusions and planets, with a remark on non-standard Dirichlet forms and polymers. In 'Stochastic Analysis and Applications', Proceedings, Swansea 1983, ed. A. Truman, 1-25. Lecture Notes in Maths. 1095. Springer-Verlag.
2. Baxendale, P. (1984) Brownian motions in the diffeomorphism group I. Compositio Math., 53, 19-50.
3. Carlen, E. (1984). Conservative Diffusions. Commun. Math. Phys., 94, 273-296.
4. Carlen, E. (1985). Potential scattering in stochastic mechanics. Ann. Inst. H. Poincaré, 42, no.4, 407-28.
5. Carverhill, A.P. (1985). Flows of stochastic dynamical systems: Ergodic Theory. Stochastics 14, 273-317.
6. Carverhill, A.P. (1985). A formula for the Lyapunov numbers of a stochastic flow. Application to a perturbation theorem. Stochastics, 14, 209-226.
7. Carverhill, A.P., Chappell, M.J. and Elworthy, K.D. (1986). Characteristic exponents for stochastic flows. In Stochastic Processes - Mathematics and Physics. Proceedings, Bielefeld 1984. Ed. S. Albeverio et al. pp.52-72. Lecture Notes in Mathematics 1158. Springer-Verlag.
8. Carverhill, A.P. and Elworthy, K.D. (1983). Flows of stochastic dynamical systems: the functional analytic approach. Z für Wahrscheinlichkeitstheorie. 65, 245-267.
9. Chappell, M.J. Lyapunov exponents for certain stochastic flows. Warwick University Ph.D. thesis (in preparation).
10. Crauel, H. (1986). Lyapunov exponents and invariant measures of stochastic systems on manifolds. In Lyapunov Exponents, Proceedings Bremen 1984, ed. L. Arnold and V. Wihstutz, pp.271-291. Lecture Notes in Maths. 1186. Springer-Verlag.
11. Elworthy, K.D. (1982). Stochastic differential equations on manifolds. London Mathematical Society Lecture Notes 70. Cambridge: Cambridge University Press.
12. Kunita, H. (1981). On the decomposition of solutions of stochastic differential equations. In Stochastic Integrals, ed. D. Williams, pp.213-255. Lecture Notes

in Maths. 851. Berlin, Heidelberg, New York: Springer-Verlag.

13. Lewis, J.T. and Truman, A. (1986). The stochastic mechanics of the ground-state of the hydrogen atom. In Stochastic Processes - Mathematics and Physics, proceedings, Bielefeld 1984. ed. S. Albeverio et al. pp.168-179. Lecture Notes in Maths, 1158, Springer-Verlag.
14. Nelson, E. (1967). Dynamical Theories of Brownian Motion. Mathematical Notes. Princeton: Princeton University Press.
15. Ruelle, D. (1984). Characteristic Exponents for a viscous fluid subjected to time dependent forces. Commun. Math. Phys. 93, 285-300 (1984).
16. Shucker, D.S. (1980) Stochastic mechanics of systems with zero potential. J. of Functional Analysis, 38, 146-155.
17. Simon, B. (1982). Schrödinger Semigroups. Bull. Amer. Math. Soc., 7, no.3, 447-526.
18. Truman, A. (1986). An introduction to the stochastic mechanics of stationary states with applications. In "From local times to global geometry, control and physics", ed. K.D. Elworthy, 329-344. Pitman Research Notes in Maths. Series, 150. Longman Scientific and Technical.
19. Walters, P. (1975). Ergodic Theory - Introductory Lectures. Lecture Notes in Maths, 458. Springer-Verlag.
20. Zheng, W. (1985). Tightness results for laws of diffusion processes, application to stochastic mechanics. Ann. Inst. H. Poincaré, 21, no.2, 103-124.

PLANETESIMAL DIFFUSIONS

by

Richard Durran and Aubrey Truman
Department of Mathematics and Computer Science
University College of Swansea
Singleton Park
Swansea SA2 8PP

Introduction

In this paper we discuss two related problems from mathematical physics. The first of these is concerned with the correspondence limit of the $\psi_{n,n-1,n-1}$ orbital of the hydrogen atom. According to the physics folklore, this limit should correspond to a Keplerian circular orbit. Here we discuss this limit in the context of Nelson's stochastic mechanics [8], [9]. We show that in the correspondence limit the sample paths of Nelson's stochastic mechanics for the $\psi_{n,n-1,n-1}$ orbital converge in an L^2 sense to a classical trajectory corresponding to a Keplerian circular orbit.

The second problem which we discuss goes back at least as far as René Descartes. This is the problem of giving a mathematical model for the condensation of planets out of a protosolar nebula [10]. Albeverio, Hoegh-Krohn and Blanchard have already discussed this problem in the setting of stochastic mechanics [1], [2]. Our treatment owes a great deal to them. These authors give a possible mechanism in stochastic mechanics for explaining the origin of the Titius-Bode law: that the successive distances of the planets to the sun are the terms in a geometric series. Our main objective in the present work is to explain how stochastic mechanics leads to coplanar, circular orbits, being described at a rate consistent with Kepler's third law of planetary motion: in suitable units the square of the periodic time is equal to the cube of the radius of the circular orbit.

We give only outline proofs in this paper. For further details the reader is advised to consult the original references [8], [9] and [4]. It is a pleasure to thank David Williams for helpful conversations.

Stochastic mechanics of the $\psi_{n,n-1,n-1}$ orbital

We consider the stochastic mechanics corresponding to a stationary state solution Ψ of the Schrödinger equation

$$i\hbar\frac{\partial\Psi}{\partial t}(\underset{\sim}{x},t) = H\Psi(\underset{\sim}{x},t),$$

where H is the Kepler-Coulomb Hamiltonian $H = -\frac{\hbar^2}{2m}\Delta_x - \frac{Ze^2}{|\underset{\sim}{x}|}$. The potential $-\frac{Ze^2}{|\underset{\sim}{x}|}$

corresponds to an inverse square law of force arising from the Coulomb attraction between a charge Ze at the origin and a charge -e carried by the orbiting particle with mass m. Needless to say $\hbar$ is Planck's constant divided by 2π.

In this paper we take $\Psi(\underset{\sim}{x},t) = \psi_{n,n-1,n-1}(\underset{\sim}{x})e^{-iE_n t/\hbar}$ where $E_n = \left\{-\frac{Ze^2}{2an^2}\right\}$ and

$$\psi_{n,n-1,n-1}(\underset{\sim}{x}) = (r\sin\theta)^{n-1}\exp\{i(n-1)\phi - \frac{r}{na}\} = (x+iy)^{n-1}\exp\left\{-\frac{(x^2+y^2+z^2)^{\frac{1}{2}}}{na}\right\},$$

$a = \hbar^2/me^2Z$ being the Bohr radius, $\underset{\sim}{x} = (x,y,z)$ in cartesians, having spherical polar coordinates (r,θ,ϕ).

The physical significance of Ψ is contained in the following eigenvalue relations:

$$(H - E_n)\psi_{n,n-1,n-1} = 0,\quad (\underset{\sim}{L}^2 - n(n-1)\hbar^2)\psi_{n,n-1,n-1} = 0,\quad (L_3 - (n-1)\hbar)\psi_{n,n-1,n-1} = 0,$$

where $\underset{\sim}{L}$ is the quantum mechanical angular momentum operator

$\underset{\sim}{L} = i^{-1}\hbar(\underset{\sim}{x}\wedge\underset{\sim}{\nabla}_x) = (L_1,L_2,L_3)$ in cartesians and $\underset{\sim}{L}^2 = L_1^2 + L_2^2 + L_3^2$.

Following Nelson [8], [9], we write $\Psi = e^{R+iS}$, where R and S are real. Then the corresponding stochastic mechanics arises from the diffusion

$$d\underset{\sim}{X}(t) = \frac{\hbar}{m}\underset{\sim}{b}(\underset{\sim}{X}(t))dt + (\frac{\hbar}{m})^{\frac{1}{2}}d\underset{\sim}{B}(t),$$

where $\underset{\sim}{b} = \underset{\sim}{\nabla}(R+S)$, $\underset{\sim}{B}$ being a $BM(\mathbb{R}^3)$ process with $\mathbb{E}\{B_i(t)B_j(s)\} = \delta_{ij}\min(s,t)$, $i,j = 1,2,3$, $\underset{\sim}{B} = (B_1,B_2,B_3)$ in cartesians.

Nelson showed that $\underset{\sim}{b}\pm$, the forward and backward drifts defined by

$$\lim_{h\downarrow 0}\mathbb{E}\left\{\frac{\underset{\sim}{X}(t\pm h) - \underset{\sim}{X}(t)}{\pm h}\middle|\underset{\sim}{X}(t)\right\} = \underset{\sim}{b}\pm(\underset{\sim}{X}(t)),$$

are given by

$$\underset{\sim}{b}\pm = \frac{\hbar}{m}(\underset{\sim}{\nabla}S \pm \underset{\sim}{\nabla}R).$$

Moreover, given sufficient regularity, defining the mean forward and backward derivatives $D\pm$ by

$$D\pm f(\underset{\sim}{X}(t),t) = \lim_{h\downarrow 0}\mathbb{E}\left\{\frac{f(\underset{\sim}{X}(t\pm h),t\pm h) - f(\underset{\sim}{X}(t)t)}{\pm h}\middle|\underset{\sim}{X}(t)\right\},$$

Nelson obtained

$$D\pm f(\underset{\sim}{X}(t),t) = (\frac{\partial}{\partial t} + b\pm.\nabla\pm\frac{\hbar}{2m}\Delta)f(\underset{\sim}{X}(t),t).$$

A tedious calculation then yields

$$2^{-1}m(D_+D_- + D_-D_+)\underset{\sim}{X}(t) = \hbar\underset{\sim}{\nabla}\frac{\partial S}{\partial t} - \frac{\hbar^2}{2m}\underset{\sim}{\nabla}\{|\underset{\sim}{\nabla}R|^2 - |\underset{\sim}{\nabla}S|^2 + \Delta R\},$$

giving by virtue of the above Schrödinger equation

$$2^{-1}m(D_+D_- + D_-D_+)\underset{\sim}{X}(t) = -\frac{Ze^2\underset{\sim}{X}(t)}{|\underset{\sim}{X}(t)|^3}.$$

This is the Nelson-Newton law for the inverse square law of force i.e. a stochastic version of Newton's second law of motion. (See refs. [8], [9] and [11]).

Setting

$$H(\underset{\sim}{X}(t)) = 2^{-1}m\left\{\frac{(\underset{\sim}{b}_+)^2(\underset{\sim}{X}(t)) + (\underset{\sim}{b}_-)^2(\underset{\sim}{X}(t))}{2}\right\} - \frac{Ze^2}{|\underset{\sim}{X}(t)|}$$

the total energy of the stochastic process $\underset{\sim}{X}$, obtained by averaging the kinetic energies due to the forward and backward drifts, and

$$\underset{\sim}{L}(\underset{\sim}{X}(t)) = 2^{-1} m \underset{\sim}{X}(t) \wedge (\underset{\sim}{b}_+(\underset{\sim}{X}(t)) + \underset{\sim}{b}_-(\underset{\sim}{X}(t))),$$

the stochastic angular momentum, we see that if the initial density distribution is the invariant one, e^{2R} in this stationary state case, then

$$\mathbb{E}\{H(\underset{\sim}{X}(t))\} = E_n, \ \mathbb{E}\{L_3(\underset{\sim}{X}(t))\} = (n-1)\hbar, \ \mathbb{E}\{(L_1^2 + L_2^2 + L_3^2)(\underset{\sim}{X}(t))\} = (n-1)(n-\tfrac{1}{2})\hbar^2.$$

These are the analogues of the above eigenvalue relations. What is more important from our point of view is that, <u>irrespective of the initial conditions</u> for the diffusion, with probability one, the forward angular momentum

$$L_3^+(\underset{\sim}{X}(t)) = (n-1)\hbar.$$

This will be crucial for the physical interpretation of $\underset{\sim}{X}$.

According to the physics folklore, the $\psi_{n,n-1,n-1}$ orbital corresponds to a circular Keplerian orbit in the correspondence limit. Thus, we expect that in the correspondence limit the Nelson diffusion should converge to a Keplerian circular orbit. Setting $\varepsilon^2 = \hbar/m$, $\lambda = \hbar n$ and $\mu = Ze^2$, the Nelson diffusion $\underset{\sim}{X}$ satisfies

$$\left.\begin{aligned} dX &= \left(\frac{\lambda}{m} - \varepsilon^2\right)\frac{(X-Y)}{(X^2+Y^2)}dt - \frac{\mu}{\lambda}\frac{X\,dt}{(X^2+Y^2+Z^2)^{\frac{1}{2}}} + \varepsilon dB_x(t), \\ dY &= \left(\frac{\lambda}{m} - \varepsilon^2\right)\frac{(X+Y)}{(X^2+Y^2)}dt - \frac{\mu}{\lambda}\frac{Y\,dt}{(X^2+Y^2+Z^2)^{\frac{1}{2}}} + \varepsilon dB_y(t), \\ dZ &= -\frac{\mu}{\lambda}\frac{Z\,dt}{(X^2+Y^2+Z^2)^{\frac{1}{2}}} + \varepsilon dB_z(t), \end{aligned}\right\} (s)$$

where (X,Y,Z) are the cartesian coordinates of $\underset{\sim}{X}$. The correspondence limit is one in which λ is fixed and ε tends to zero. We shall discuss this in the next section but before doing this we investigate the underlying deterministic equations.

We rewrite the above Itô equations in polar coordinates giving

$$\left.\begin{aligned} dR &= \frac{\lambda}{m}\left(\frac{1}{R} - \frac{m\mu}{\lambda^2}\right)dt + \varepsilon d\tilde{B}_R(t), \\ d\theta &= \left(\frac{\lambda}{m} - \frac{\varepsilon^2}{2}\right)\frac{\cot\theta}{R^2}dt + \frac{\varepsilon}{R}d\tilde{B}_\theta(t), \\ d\phi &= \left(\frac{\lambda}{m} - \varepsilon^2\right)\frac{\operatorname{cosec}^2\theta}{R^2}dt + \frac{\operatorname{cosec}\theta}{R}\varepsilon d\tilde{B}_\phi(t). \end{aligned}\right\} (s)$$

The classical limit formally corresponds to

$$\dot{R} = \frac{\lambda}{m}\left(\frac{1}{R} - \frac{m\mu}{\lambda^2}\right), \ \dot{\theta} = \frac{\lambda}{m}\frac{\cot\theta}{R^2}, \ \dot{\phi} = \frac{\lambda}{m}\frac{\operatorname{cosec}^2\theta}{R^2}. \qquad (c)$$

Evidently there is a circular orbit K at $\theta = \pi/2$ and $R = \lambda^2/m\mu$, K being described with angular speed $\dot{\phi} = \lambda/mR^2 = m\mu^2/\lambda^3$. K is the Keplerian circular orbit for the above inverse square law of force. We begin with an elementary proposition.

<u>Proposition 0</u>

The circular orbit K is the stable limit cycle for the correspondence limit (c).

Outline proof

Work in units with $\frac{\lambda}{m} = \frac{\mu}{\lambda} = 1$. Then for $R(0) \neq 1$ we obtain

$$- R(t) + R(0) - \ln \frac{1 - R(t)}{1 - R(0)} = t,$$

so $\frac{d^2 t}{dR^2} > 0$. Hence, (R,t) graph must be convex upwards giving the (t,R) graph below

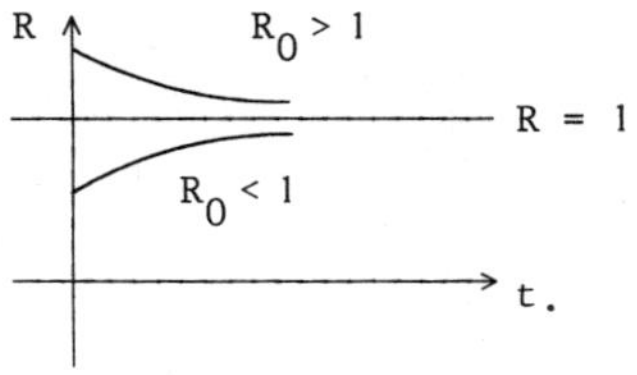

Hence, we see that $R(t) \to 1$ as $t \to \infty$, so $T(t) = \int_0^t \frac{ds}{R^2(s)} \uparrow \infty$ as $t \uparrow \infty$ and

$$\cos\theta(t) = e^{-T(t)} \cos\theta_0 \to 0 \quad \text{as} \quad t \uparrow \infty .$$

Also $\dot{\phi}(t) \to 1$ as $t \uparrow \infty$, proving the desired result. //

The correspondence limit of the $\psi_{n,n-1,n-1}$ orbital

Here we take λ to be a fixed quantity and consider the limiting case as ε tends to zero of the solution $\underset{\sim}{X} = \underset{\sim}{X}^{\varepsilon}$ to the above stochastic differential equation (s). This is the correspondence limit according to Bohr. The methods of Carlen [3] and Zheng [13] guarantee the existence of solutions to the above stochastic differential equation. There is a simple a priori estimate for $R^{\varepsilon} = |\underset{\sim}{X}^{\varepsilon}|$ obtained by comparing R^{ε} with $R = |\underset{\sim}{X}^{o}|$, the radial coordinate of the deterministic solution to equation (c), with the same initial value as R^{ε}.

Lemma 1

For each finite $t \geq 0$

$$\mathbb{E}\{(R^{\varepsilon} - R)^2(t)\} \leq \int_0^t \varepsilon^2 \, du.$$

Proof

Define S by

$$S(t) = (R^{\varepsilon} - R)^2(t) - \int_0^t \varepsilon^2 \, du + \varepsilon^2 T,$$

where T is a constant with $T > t$. Then Itô's formula shows that

$$dS(t) = -2\lambda (R^{\varepsilon} - R)^2(t)/R^{\varepsilon}(t)R(t)m - 2(R^{\varepsilon} - R)(t)\,\varepsilon\, d\beta(t),$$

β being a $BM(\mathbb{R})$ process. Since the first term is negative semidefinite S is a local supermartingale. But S is a positive process, and by Fatou's lemma a positive local supermartingale is a true supermartingale [12]. Hence, for $u < t$,

$$\mathbb{E}[S(t)] = \mathbb{E}\{\mathbb{E}[S(t)|\mathcal{F}_u]\} \le \mathbb{E}[S(u)],$$

$\mathcal{F}_t$ being the usual filtration. Letting $u \to 0$ gives the desired result. //

Important remark on uniform convergence in probability

Doob's supermartingale inequality applied to $S(t) = (R^\varepsilon - R)^2(t) + \int_t^u \varepsilon^2 ds$, for $t < u$, yields

$$\lambda\ \mathbb{P}[\sup_{t \le u} (R^\varepsilon - R)^2(t) > \lambda] \le \int_0^u \varepsilon^2 ds.$$

In other words $R^\varepsilon(\cdot) \to R(\cdot)$ uniformly in probability on compact intervals. Letting $\varepsilon = \varepsilon_n$, with $\sum \lambda_n^{-1} \varepsilon_n^2 < \infty$ for $\lambda = \lambda_n \to 0$, an application of the first Borel Cantelli lemma shows that almost surely $R^\varepsilon(\cdot) \to R(\cdot)$ uniformly on compact intervals. Also, when ε is a function of time with $\int_0^\infty \varepsilon^2(s)\, ds < \infty$, Doob's supermartingale convergence theorem gives $\lim_{t \uparrow \infty} (R^\varepsilon(t) - R(t))^2$ exists almost surely.

In any case we have proved:

Proposition 2

Let $R^\varepsilon(t)$ satisfy

$$dR^\varepsilon(t) = \frac{\lambda}{m}\left(\frac{1}{R^\varepsilon(t)} - \frac{m\mu}{\lambda^2}\right) dt + \varepsilon d\beta(t),$$

β being a BM($\mathbb{R}$) process. Define $R(t)$ by

$$\frac{dR(t)}{dt} = \frac{\lambda}{m}\left(\frac{1}{R(t)} - \frac{m\mu}{\lambda^2}\right),$$

where it is understood that R^ε and R have the same initial values. Then

$$\mathbb{E}(R^\varepsilon - R)^2(t) \le \varepsilon^2 t.$$

Writing $\underset{\sim}{X}^\varepsilon = (X^\varepsilon, Y^\varepsilon, Z^\varepsilon)$, recall that Z^ε satisfies

$$dZ^\varepsilon = -\frac{\mu}{\lambda}\frac{Z^\varepsilon}{R^\varepsilon} dt + \varepsilon\, d\beta_z(t).$$

There is a corresponding proposition for Z^ε.

Proposition 3

With R defined as above, let Z satisfy

$$\frac{dZ(t)}{dt} = -\frac{\mu}{\lambda}\frac{Z(t)}{R(t)},$$

Z having the same initial value as Z^ε. Then

$$\mathbb{E}(Z^\varepsilon - Z)^2(t) \le \frac{\mu}{\lambda}\int_0^t \frac{\mathbb{E}(R^\varepsilon - R)^2(s)}{R(s)}\,ds + \varepsilon^2 t.$$

Outline proof

Itô's formula gives after some calculation:

$$d(Z^\varepsilon - Z)^2 = -\frac{2\mu}{\lambda}\frac{(Z^\varepsilon - Z)^2}{R}dt + \frac{2\mu}{\lambda}(Z^\varepsilon - Z)Z^\varepsilon\left(\frac{1}{R} - \frac{1}{R^\varepsilon}\right)dt + \varepsilon^2 dt + 2(Z^\varepsilon - Z)\,\varepsilon\, d\beta_Z.$$

To deal with the second term we use the Cauchy Schwarz inequality giving

$$\mathbb{E}\left\{\frac{Z^\varepsilon(Z^\varepsilon - Z)(R^\varepsilon - R)}{RR^\varepsilon}\right\} \le \frac{1}{R}\,\mathbb{E}^{\frac{1}{2}}(R^\varepsilon - R)^2\,\mathbb{E}^{\frac{1}{2}}(Z^\varepsilon - Z)^2\left(\frac{Z^\varepsilon}{R^\varepsilon}\right)^2 \le \frac{\mathbb{E}^{\frac{1}{2}}(R^\varepsilon - R)^2\,\mathbb{E}^{\frac{1}{2}}(Z^\varepsilon - Z)^2}{R}$$

and, since $xy \le 2^{-1}(x^2 + y^2)$, we obtain

$$\mathbb{E}\left\{Z^\varepsilon(Z^\varepsilon - Z)\left(\frac{1}{R} - \frac{1}{R^\varepsilon}\right)\right\} \le \frac{1}{2R}\left\{\mathbb{E}(R^\varepsilon - R)^2 + \mathbb{E}(Z^\varepsilon - Z)^2\right\}.$$

Using our first lemma, the local martingale term $\varepsilon\int_0^t (Z^\varepsilon - Z)(s)\,d\beta_Z(s)$ has zero expectation. The desired result follows taking expectation values of the above Itô formula. //

Combining the last two propositions we see that

$$\mathbb{E}(Z^\varepsilon(t) - Z(t))^2 \le \varepsilon^2 t\left(1 + \frac{\mu t}{2\lambda R_{min}}\right),$$

R_{min} being the infimum of $R(s)$ for $s \in [0,t]$. Thus Z^ε and R^ε have the correct limiting values in the correspondence limit. What about X^ε and Y^ε?

From the spherical polar form of equations (s) we obtain

$$\phi^\varepsilon(t) - \phi^\varepsilon(0) - \frac{\lambda}{m}T^o(t) = \frac{\lambda}{m}[T^\varepsilon(t) - T^o(t)] - \varepsilon^2 T^\varepsilon(t) + \varepsilon\tilde{B}(T^\varepsilon(t)),$$

T^ε being the random time

$$T^\varepsilon(t) = \int_0^t ((X^\varepsilon)^2 + (Y^\varepsilon)^2)^{-1}(s)\,ds.$$

This is the stochastic analogue of the conservation of the third component of angular momentum:

$$(X^2 + Y^2)(t)\,\dot{\phi}(t) = \frac{\lambda}{m},$$

or

$$(R^2\sin^2\theta)(t)\,\dot{\phi}(t) = \frac{\lambda}{m}.$$

We conclude this section with an elementary result consistent with Kepler's third law of planetary motion. Define the random variable $\int_t^{t+h} L_3^\varepsilon\,du$ by

$$\int_t^{t+h} L_3^\varepsilon \, du = \int_t^{t+h} (R^\varepsilon \sin \theta^\varepsilon)^2(u) \, d\phi^\varepsilon(u).$$

Then $\int_t^{t+h} L_3^\varepsilon \, du$ is the stochastic analogue of the integral with respect to the time of the third component of angular momentum.

Proposition 4

For each fixed $t, h \geq 0$,

$$\mathbb{E}\left\{\left(\int_t^{t+h} L_3^\varepsilon \, du - \frac{\lambda h}{m}\right)^2\right\} \to 0,$$

as $\varepsilon \to 0$.

Outline proof

By inspection

$$\mathbb{E}\left\{\left(\int_t^{t+h} L_3^\varepsilon \, du - \frac{\lambda h}{m}\right)^2\right\} = \mathbb{E}\left\{\left(\int_t^{t+h} \varepsilon (R^\varepsilon \sin \theta^\varepsilon)(u) \, d\beta(u) - \varepsilon^2 h\right)^2\right\},$$

β being a $BM(\mathbb{R})$ process. Therefore, we obtain

$$\lim_{\varepsilon \downarrow 0} \mathbb{E}\left\{\left(\int_t^{t+h} L_3^\varepsilon \, du - \frac{\lambda h}{m}\right)^2\right\} = \lim_{\varepsilon \downarrow 0} \int_t^{t+h} \varepsilon^2 \, \mathbb{E}(R^\varepsilon \sin \theta^\varepsilon)^2(u) \, du = 0,$$

by the above. //

Planetesimal diffusions associated with the correspondence limit

Following Albeverio et al [1], we consider a protosolar nebula with its mass concentrated near the origin. We model a situation in which the nebula is rotating about the z axis and in which a planetesimal is condensing out of the nebula. Due to collisions with the dust particles of the nebula the planetesimal accrues mass and exchanges energy and angular momentum with the nebula.

In the above we take the planetesimal to be at $\underset{\sim}{X}$ and we set μ to be unity. We take m^{-1} to be the gravitational mass of the nebula, μ_{nebula}. At time t we take the total energy per unit mass of the planetesimal to be $m^{-1} \, \mathbb{E}(H(\underset{\sim}{X}(t)))$. From the results above it is natural to interpret $(\lambda/m - \varepsilon^2)$ to be the third component of the angular momentum per unit mass of the planetesimal.

As in Albeverio et al [2], we assume that when the mass of the planetesimal, energy per unit mass and angular momentum per unit mass of the planetesimal are all constant the Nelson-Newton law is valid in the form:

$$2^{-1}(D_+D_- + D_-D_+) \, \underset{\sim}{X}(t) = -\mu_{nebula} \, \underset{\sim}{X}(t)/|\underset{\sim}{X}(t)|^3.$$

Several authors have tried to derive this law by making various statistical assumptions about the cloud but so far without success. Nevertheless the above Nelson-Newton law is derivable from the stochastic variational principle of Guerra and Morato [6].

To obtain a model in which the Nelson-Newton law is valid we use the above model with $m = m(t)$, $\lambda = \lambda(t)$. We shall assume that the diffusion coefficient $\varepsilon^2 = \hbar/m = \varepsilon^2(t) \to 0$ as $t \uparrow \infty$. We show in the next few pages that, if $\varepsilon^2(t)$ tends to zero sufficiently rapidly and if $\lambda(t)/m(t) \to L_0$ (> 0) and $\lambda^2(t)/m(t) \to r_0$ as $t \uparrow \infty$, then after a sufficiently long time the orbit of the planetesimal is in an L^2-neighbourhood of $K(r_0)$, the Keplerian circular orbit of radius r_0 (> 0).

We have not yet managed to prove that the planetesimal describes $K(r_0)$ with an angular speed consistent with Kepler's laws but it seems likely that the only extra requirement needed to ensure this is that $\mu_{nebula}(t) = m^{-1}(t) \to \mu_{sun}$, the gravitational mass of the sun, as $t \uparrow \infty$. The latter condition is necessary for the deterministic part of our equations to give the correct values of planetary years.

The proof is no harder when $\frac{\lambda}{m}(t) \to L_0$ than it is when $\frac{\lambda}{m} \equiv 1$. We shall assume in what follows that $\frac{\lambda}{m} \equiv 1$ and that $f(t) = m(t)/\lambda^2(t) \to 1, f(t) > 0$. The key to getting the correct deterministic behaviour is the next proposition.

Proposition 5

Let $\frac{dR}{dt} = \frac{1}{R(t)} - f(t)$, $f(t) > 0$, where $f(t) \to 1$ as $t \uparrow \infty$ in such a way that $\int_0^\infty |1 - f(u)|\,du < \infty$. Then $R(t) \to 1$ as $t \uparrow \infty$.

Outline proof

$$R(t) - R(0) = \int_0^t \frac{ds}{R(s)} - \int_0^t f(s)\,ds.$$

Since $R(t) \geq 0$ and $\int_0^t f(s)\,ds$ diverges to infinity we see that $\int_0^t \frac{ds}{R(s)}$ diverges to infinity also. But we can write our equation of motion in the form

$$\frac{d}{dt}(R(t) - 1) + (R(t) - 1)\frac{1}{R(t)} = (1 - f(t)),$$

giving

$$e^{\int_0^t \frac{ds}{R(s)}}(R(t) - 1) = R(0) - 1 + \int_0^t (1 - f(u))\, e^{\int_0^u \frac{ds}{R(s)}}\,du.$$

Hence,

$$|R(t) - 1| \le |R(0) - 1| + \int_0^t |1 - f(u)|\, du < \infty.$$

so that $A = \overline{\lim_{t\uparrow\infty}}\, R(t) < \infty$. Since $\int_0^t \frac{ds}{R(s)}$ diverges to infinity, de l'Hôpital's rule gives

$$\overline{\lim_{t\uparrow\infty}}\, |R(t) - 1| \le A\, \overline{\lim_{t\uparrow\infty}}\, |(1 - f(t))| = 0. \quad //$$

The corresponding stochastic equation which we have to consider is

$$dR^{\varepsilon}(t) = \left(\frac{1}{R^{\varepsilon}(t)} - f(t)\right)dt + \varepsilon(t)d\beta(t).$$

Proposition 6

If $\int_0^{\infty} \varepsilon^2(u)\, du < \infty$, then $\lim_{t\uparrow\infty} \mathbb{E}\{(R^{\varepsilon} - R)^2(t)\}$ exists.

Proof

Itô's formula gives

$$\frac{d}{dt}\, \mathbb{E}\{(R^{\varepsilon} - R)^2(t)\} = -2\, \mathbb{E}(R^{\varepsilon} - R)^2(R^{\varepsilon})^{-1} R^{-1}(t) + \varepsilon^2(t) \le \varepsilon^2(t).$$

The desired result now follows from Cauchy's generalised convergence criterion. //

It follows that $\mathbb{E}\{(R^{\varepsilon} - R)^2(t)\} \to \ell \ge 0$ as $t \uparrow \infty$. We now show that $\ell = 0$. We require a further lemma:

Lemma 7

For each $t \ge 0$

$$\int_0^t \mathbb{E}\{((R^{\varepsilon})^2 - R^2)^2\, (R^{\varepsilon})^2(u)\}\varepsilon^2(u)\, du < \infty,$$

if $\int_0^{\infty} \varepsilon^2(u)\, du < \infty$.

Proof

This is a simple bootstrap based on the remark after the first lemma [7]. Consider

$$d\{(R^{\varepsilon} - R)^4(t) + 6\int_t^{\infty} (R^{\varepsilon} - R)^2(u)\varepsilon^2(u)du\} = 4(R^{\varepsilon} - R)^3 \left(\frac{1}{R^{\varepsilon}} - \frac{1}{R}\right)dt + 4(R^{\varepsilon} - R)^3\, \varepsilon(t)d\beta(t),$$

β being a BM($\mathbb{R}$) process. Observe that, as in the first lemma, the first term is negative semidefinite. Then, as before,

$$\{(R^\varepsilon - R)^4(t) + 6\int_t^\infty (R^\varepsilon - R)^2(u)\varepsilon^2(u)du\}$$

is a true supermartingale giving

$$\mathbb{E}(R^\varepsilon - R)^4(t) \le 6\,\mathbb{E}\int_0^t (R^\varepsilon - R)^2(u)\varepsilon^2(u)du \le 6\int_0^\infty \varepsilon^2(u)du\int_0^t \varepsilon^2(u)du < \infty.$$

Therefore, by Doob's martingale convergence theorem, $\lim(R^\varepsilon - R)^4(t)$ exists almost surely. Now consider

$$d\{(R^\varepsilon - R)^6(t) + 15\int_t^\infty (R^\varepsilon - R)^4(u)\varepsilon^2(u)du\} = 6(R^\varepsilon - R)^5\left(\frac{1}{R^\varepsilon} - \frac{1}{R}\right)dt + 6(R^\varepsilon - R)^5\varepsilon(t)d\beta(t)$$

and repeat the last argument giving

$$\mathbb{E}(R^\varepsilon - R)^6(t) < \infty. \quad //$$

<u>Proposition 8</u>

If $\int_0^\infty \varepsilon^2(u)du < \infty$ then $\mathbb{E}(R^\varepsilon - R)^2(t) \to 0$ as $t \uparrow \infty$.

<u>Proof</u>

This uses the following nice coincidence of signs. Itô gives

$$d\{(R^\varepsilon)^2 - R^2\}^2 = -4f(R^\varepsilon - R)^2(R^\varepsilon + R)dt + 6(R^\varepsilon)^2\varepsilon^2dt - 2\varepsilon^2R^2dt + 4((R^\varepsilon)^2 - R^2)R^\varepsilon\,\varepsilon d\beta,$$

β being a BM($\mathbb{R}$) process. From the last lemma the final term has zero expectation giving, since the third term is negative semidefinite,

$$0 \le \mathbb{E}((R^\varepsilon)^2 - R^2)^2(t) \le -4\int_0^t f(s)R(s)\,\mathbb{E}(R^\varepsilon - R)^2(s)ds - 4\int_0^t f(s)\,\mathbb{E}(R^\varepsilon - R)^2(s)R^\varepsilon(s)ds$$
$$+ 6\int_0^t \varepsilon^2(s)\,\mathbb{E}(R^\varepsilon)^2(s)ds.$$

Since the second term is negative semidefinite, we obtain from the remark after the first lemma

$$\int_0^t f(s)R(s)\,\mathbb{E}(R^\varepsilon - R)^2(s)ds \le \frac{3}{2}\int_0^t \varepsilon^2(s)\,\mathbb{E}(R^\varepsilon)^2(s)ds < \infty.$$

Since $f(s) \to 1$ and $R(s) \to 1$ as $s \uparrow \infty$, we obtain

$$\int_0^\infty \mathbb{E}(R^\varepsilon - R)^2(s)ds < \infty.$$

Combining this estimate with the last proposition proves the result. //

The Z convergence can be handled in much the same way as before.

Proposition 9

If $\int_0^\infty \varepsilon^2(u)\,du < \infty$, then $\mathbb{E}(Z^\varepsilon - Z)^2(t) \to 0$ as $t \uparrow \infty$.

Outline proof

Reworking the argument which we used to obtain Bohr's correspondence limit gives

$$\frac{d}{dt}\,\mathbb{E}(Z^\varepsilon - Z)^2(t) \le -\frac{1}{\lambda(t)R(t)}\,\mathbb{E}(Z^\varepsilon - Z)^2(t) + e(t),$$

where the function $e(t)$ is given by

$$e(t) = \mathbb{E}(R^\varepsilon - R)^2(t)/\lambda(t)R(t) + \varepsilon^2(t).$$

Rearranging as before, we obtain

$$e^{\int_0^t \frac{ds}{\lambda(s)R(s)}}\,\mathbb{E}(Z^\varepsilon - Z)^2(t) \le \int_0^t e(s)\, e^{\int_0^s \frac{du}{\lambda(u)R(u)}}\,ds.$$

De l'Hôpital gives

$$\overline{\lim_{t\uparrow\infty}}\,\mathbb{E}(Z^\varepsilon - Z)^2(t) \le \lim_{t\uparrow\infty} e(t)\,\lambda(t)R(t) = 0$$

by the last proposition, if $\int_0^\infty \varepsilon^2(u)\,du < \infty$. //

We therefore see that, if $\int_0^\infty \varepsilon^2(u)\,du < \infty$, ε^2 being the diffusion coefficient, and if $L(t)$, the third component of angular momentum per unit mass of the planetesimal, is such that $L(t) \to L_0$ (>0) as $t \uparrow \infty$, as long as $r_0 > 0$ is defined by $\mu_{nebula}(t) \to L_0^2/r_0 = \mu_{nebula}$, the stochastic process is, after a sufficiently long time, in an L^2 neighbourhood of $K(r_0)$, the Keplerian circular orbit of radius r_0. We hope to show in a future publication that, if only $\mu_{nebula} = \mu_{sun}$, the gravitational mass of the sun, then after infinite time $K(r_0)$ is being described with an angular speed consistent with the observed duration of the planetary years. We conclude with a simple result in this direction:

Proposition 10

Let ε be a function of time with $\int_0^\infty \varepsilon^2(u)\,du < \infty$. Using an obvious extension of the notation of the last section, with ε a function of time, for fixed $h \geq 0$,

$$\mathbb{E}\left\{\left(\int_t^{t+h} L_3^{\varepsilon}\, du - \int_t^{t+h} \frac{\lambda}{m}(u)\, du\right)^2\right\} \to 0$$

as $t \uparrow \infty$.

Proof

Reinstating λ and m to emphasise the physical significance of the result, the proof is a trivial extension of the proof of proposition 4. //

REFERENCES

[1] Albeverio, S., Blanchard, Ph., Hoegh-Krohn, R.(1984): A stochastic model for the orbits of planets and satellites: an interpretation of the Titius-Bode law. Expositiones Mathematicae 1, 365-373.

[2] Albeverio, S., Blanchard, Ph., Hoegh-Krohn, R: Newtonian Diffusions and Planets, with a remark on non-standard Dirichlet forms and Polymers. In 'Stochastic Analysis and Applications', Proceedings, Swansea 1983, editors A. Truman and D. Williams, 1-25. Lecture Notes in Maths. 1095, Springer Verlag.

[3] Carlen, E. (1984): Conservative Diffusions. Commun. Math. Phys., 94, 273-296.

[4] Durran, R.M., and Truman, A: Planetesimal Diffusions and Stochastic Mechanics, in preparation.

[5] Gihman, I.I., and Skorohod, A.V. (1972): Stochastic Differential Equations, Ergebnisse der Mathematik. Berlin: Springer Verlag.

[6] Guerra, F., and Morato, L.M. (1983): Quantization of dynamical systems and stochastic control theory. Phys. Rev. D, 1774-1786.

[7] McKean, H.P. (1969): Stochastic Integrals. Probability and Mathematical Statistics Monographs. New York: Academic Press.

[8] Nelson, E. (1967): Dynamical Theories of Brownian Motion. Mathematical Notes. Princeton: Princeton University Press.

[9] Nelson, E. (1985): Quantum Fluctuations. Princeton Series in Physics. Princeton: Princeton University Press.

[10] Nieto, M.M. (1972): The Titius Bode Law of Planetary Distances, its History and Theory. Oxford: Pergamon Press.

[11] Truman, A. (1986): An introduction to the stochastic mechanics of stationary states with applications. In 'From local times to global geometry, control and physics', editor K.D. Elworthy, 329-344. Pitman Research Notes in Maths. Series 150. Longman Scientific and Technical.

[12] Williams, D. (1979): Diffusions, Markov Processes and Martingales. Volume 1. Foundations; and, jointly with Rogers, L.C.G. (1987) Volume 2. Chichester: John Wiley.

[13] Zheng, W.A. (1985): Tightness results for laws of diffusion processes and applications to stochastic mechanics. Ann. Inst. Henri Poincaré 21, #2, 103-124, and references cited therein.

Footnotes added in proof

1. We are grateful to Professor W.R. Schneider for a helpful remark correcting our expression for $\mathbb{E}\{\underset{\sim}{L}^2\}$ here. The discrepancies between the stochastic and quantum mechanical averages disappear altogether if we work with $\underset{\sim}{b}_+$ by itself. (See ref (4)).

2. After this paper was typed we received a very interesting preprint from Professors Piotr Garbaczewski and Dariusz Prorok, 'Semiclassical quantum mechanics for the Coulomb-Kepler problem', discussing the same problem by very different means. It would seem likely that more progress could be made by combining the methods of their paper with ours.

BROWNIAN MOTION ON HYPERSURFACES AND COMPUTER SIMULATION

by

Richard M. Durran and Aubrey Truman
Department of Mathematics and Computer Science
University College of Swansea
Singleton Park, SWANSEA, SA2 8PP

1. Introduction

One of the motivations for this work was to construct a simple and accurate computer simulation of Brownian motion on a surface embedded in $\mathbb{R}^3$. To this end, we shall discuss a characterization of Brownian motion on hypersurfaces in $\mathbb{R}^d$, in terms of local (parametric) coordinate processes. We shall obtain, via a straight forward application of Itô's formula and some elementary differential geometry expressions for the stochastic differentials of these processes. We go on to discuss the relation between this characterization, and those given by Baxendale [2], Lewis [6] and, for Brownian motion on surfaces in $\mathbb{R}^3$, by Williams [7]. For Brownian motion on surfaces in $\mathbb{R}^3$ we give an especially simple formula for the stochastic differentials of local coordinates. Using this formula we give a few examples of Brownian motion on surfaces in $\mathbb{R}^3$ and discuss, in detail, their computer simulation.

Finally, as the emphasis of this work is on stochastic mechanics and stochastic processes, we discuss briefly some simulations of the stochastic processes arising from the stochastic mechanics of the first few excited states of the hydrogen atom. The coloured pictures which one can obtain here could give quite detailed information about the charge distributions for the excited states of hydrogen. Unfortunately it is only possible to give black and white reproductions in this paper. We have used a screen dump program with shading to give some impression of colour.

2. Brownian Motion on Hypersurfaces

Let V be a hypersurface embedded in $\mathbb{R}^d$ with orientating normal vector field $n = n(x) \in S^{d-1}$ (the unit sphere in $\mathbb{R}^d$) at each point $x \in V$. Let P be the projection onto the tangent plane, T_x, to the hypersurface at x. Using the summation convention define, for $f : V \to \mathbb{R}$, $(P\nabla)$ by

$$(P\nabla f)_i(x) = (\delta_{ij} - n_i n_j)\frac{\partial f}{\partial x_j}(x), \qquad i = 1,2,\ldots,d, \tag{2.1}$$

$n = (n_1, n_2, \ldots, n_d)$ being the unit normal in cartesian coordinates.

Let $u_1, u_2, \ldots, u_d$ be a local coordinate system, with $u_1, u_2, \ldots, u_{d-1}$ the local parametric coordinates of the hypersurface, so that each point $x = x(u) = x(u_1, u_2, \ldots, u_{d-1})$. The tangent vectors $\frac{\partial x}{\partial u_1}(u)$, $\frac{\partial x}{\partial u_2}(u)$, ..., $\frac{\partial x}{\partial u_{d-1}}(u)$

are linearly independent and span $T_{x(u)}$ by assumption.

The Laplacian Δ on the surface is defined (writing $P\nabla = \nabla_p$) by

$$\Delta f = \text{Trace}\,(\nabla_p \nabla_p f)\,. \tag{2.2}$$

The all important identity in what follows is contained in the next lemma. We feel that this identity has not been given sufficient emphasis in the literature.

<u>Lemma</u>

$$\Delta x(u) = \frac{\partial x}{\partial u_i}(u)\,\Delta u_i + \frac{\partial^2 x}{\partial u_i \partial u_j}(u)\,(\nabla_p u_i . \nabla_p u_j) = (d-1)H(x)n(x)\,, \tag{2.3}$$

where $H(x)$ is the mean curvature of V at x.

<u>Proof</u>

Let $f = f(u) = f(u_1, u_2, \ldots, u_{d-1})$. Then by definition (2.1),

since $\dfrac{\partial f}{\partial x_j} = \dfrac{\partial f}{\partial u_k}\dfrac{\partial u_k}{\partial x_j}$,

$$\nabla_p f(u) = \frac{\partial f}{\partial u_k}(u)\,\nabla_p u_k\,. \tag{2.4}$$

Hence using definition (2.2) and the rule for differentiating a product we see that

$$\Delta f = \text{Trace}\,(\nabla_p \nabla_p f) = \text{Trace}\left\{\frac{\partial f}{\partial u_k}(u)\nabla_p\nabla_p u_k + (\nabla_p \frac{\partial f}{\partial u_k}(u))\nabla_p u_k\right\}.$$

$$\therefore \quad \Delta f(u) = \frac{\partial f}{\partial u_k}(u)\Delta u_k + \frac{\partial^2 f}{\partial u_i \partial u_k}(u)\,(\nabla_p u_k . \nabla_p u_i)\,. \tag{2.5}$$

But, using definitions (2.1) and (2.2) componentwise and the fact that $n.n = 1$, it is easy to check that for $x \in V$

$$\Delta x = -(\text{div}\, n(x))n(x) \tag{2.6}$$

and observing that

$$H(x) = -\frac{1}{(d-1)}\,\text{div}\, n(x)\,, \tag{2.7}$$

where $H(x)$ is the mean curvature of V at x we see that

$$\Delta x = (d-1)H(x)n(x)\,. \tag{2.8}$$

Finally, applying (2.5) componentwise to x and using (2.8), identity (2.3) follows easily. //

We shall use this identity in the proof of the following proposition, but first we note a simple corollary originally due to Lewis [6].

<u>Corollary</u>

Denote by BM(V) (a Brownian motion on V) a process on V with generator

$2^{-1}\Delta$. The BM(V) process on the hypersurface V is a martingale in the ambient Euclidean space if and only if the mean curvature of V vanishes identically.

Proof

We merely observe that the vanishing of the drift $2^{-1}\Delta X$ gives precisely this by virtue of (2.8). //

Proposition

Let $B(t)$ be a $BM(\mathbb{R}^d)$, i.e. a Brownian motion on $\mathbb{R}^d$. Then a Brownian motion, $X(t) = X(u(t)) = X(u_1(t), u_2(t), \ldots, u_{d-1}(t))$, on a hypersurface V embedded in $\mathbb{R}^d$ is characterized by the stochastic differentials

$$du_i(t) = \frac{1}{2}\Delta u_i\, dt + (R\nabla_p u_i).dB(t)\ , \qquad i = 1,2,\ldots,(d-1)\ , \tag{2.9}$$

of the (d-1) parametric coordinate processes, $R(\in O(d))$ being any non-anticipating orthogonal transformation.

Proof

Let $X = X(u) = X(u_1, u_2, \ldots, u_{d-1}) \in V$ and set

$$du_i(t) = b_i dt + d\beta_i(t)\ , \qquad i = 1,2,\ldots,(d-1)\ , \tag{2.10}$$

a set of (d-1) Itô equations. We try to choose b_i and $d\beta_i(t)$ so that $X(t) = X(u(t)) = X(u_1(t), u_2(t), \ldots, u_{d-1}(t))$ is a Brownian motion on the hypersurface V.

Applying Itô's formula to $X(u(t))$ gives

$$dX(t) = \frac{\partial X}{\partial u_i}(u)\ du_i(t) + \frac{1}{2}\frac{\partial^2 X}{\partial u_i \partial u_j}(u)\ du_i(t)\ du_j(t)$$

$$= \frac{\partial X}{\partial u_i}(u)\ (b_i dt + d\beta_i(t)) + \frac{1}{2}\frac{\partial^2 X}{\partial u_i \partial u_j}(u)\ d\beta_i(t)\ d\beta_j(t)\ . \tag{2.11}$$

Assuming the above set of (d-1) Itô equations admits a unique solution the point $X(t)$ clearly remains on the hypersurface V. Hence for a Brownian motion on V we merely require the generator of the diffusion (2.11) to be $\frac{1}{2}\Delta$. i.e. we require

$$\frac{1}{2}\Delta X(u)\ dt = \frac{\partial X}{\partial u_i}(u)\ b_i dt + \frac{1}{2}\frac{\partial^2 X}{\partial u_i \partial u_j}(u)\ d\beta_i(t)\ d\beta_j(t)\ . \tag{2.12}$$

Now from the left hand equality of (2.3)

$$\frac{1}{2}\Delta X(u)\ dt = \frac{\partial X}{\partial u_i}(u)\ \frac{1}{2}\Delta u_i dt + \frac{1}{2}\frac{\partial^2 X}{\partial u_i \partial u_j}(u)(\nabla_p u_i . \nabla_p u_j)\ dt\ .$$

Comparing these two equations we see that for a $BM(V)$ we merely require

$$b_i = \frac{1}{2}\Delta u_i \quad \text{and} \quad d\beta_i(t)d\beta_j(t) = \nabla_p u_i . \nabla_p u_j dt\ , \quad i,j = 1,2,\ldots,d-1\ . \tag{2.13}$$

Evidently we can solve the second set of equations in (2.13) by the choice

$$d\beta_i(t) = (R\nabla_p u_i).dB(t)\ , \qquad i = 1,2,\ldots,(d-1)\ , \tag{2.14}$$

where $B(t)$ is $BM(\mathbb{R}^d)$ and $R \in O(d)$ any orthogonal transformation. //

Thus a process $X(t)$ on V satisfying,

$$dX(t) = \frac{1}{2}\Delta X(u)\ dt + \frac{\partial X}{\partial u_i}(u)\ (R\nabla_p u_i).dB(t) \tag{2.15}$$

is Brownian motion on V, with coordinate processes, $u_i(t)$, satisfying

$$du_i(t) = \frac{1}{2}\Delta u_i dt + (R\nabla_p u_i).dB(t)\ , \qquad i = 1,2,\ldots,(d-1)\ .$$

The relation between this characterization and others, is borne out by specific choices of the transformation R. Firstly let R be the identity transformation, clearly orthogonal, so that (2.15) reads, after substitution from the outer equality of (2.3),

$$dX(t) = \frac{(d-1)}{2} H(X)n(X)\ dt + \frac{\partial X}{\partial u_i}(u)\ \nabla_p u_i .dB(t)\ .$$

Noting that $\frac{\partial X}{\partial u_i}(u)\ \nabla_p u_i .dB(t) = P(X)\,dB(t)$, the projection of $dB(t)$ onto the tangent plane T_X at X, this reads,

$$dX(t) = \frac{(d-1)}{2} H(X)n(X)\ dt + P(X)\ dB(t)\ , \tag{2.16}$$

which is the characterization given by Baxendale [2], used also by Lewis [6], for Brownian motion on a hypersurface.

We now focus our attention on Brownian motion on surfaces embedded in $\mathbb{R}^3$. Let the orientating vector field $n = n(x)$ be given by

$$n(x) = \left(\frac{\partial x}{\partial u_1}(u) \wedge \frac{\partial x}{\partial u_2}(u)\right) / \left|\frac{\partial x}{\partial u_1}(u) \wedge \frac{\partial x}{\partial u_2}(u)\right|\ . \tag{2.17}$$

It is not difficult to show that

$$n(X) \wedge dB(t) = \frac{\partial X}{\partial u_i}(u)(\nabla_p u_i \wedge n(X)).dB(t)\ . \tag{2.18}$$

Thus, taking R to be a rotation of 90° about n, (2.15) reads, after substitution from (2.6) and (2.18),

$$dX(t) = -\frac{1}{2}(\mathrm{div}\ n(X))n(X)\ dt + n(X) \wedge dB(t)\ , \tag{2.19}$$

which is the characterization for Brownian motion on surfaces in $\mathbb{R}^3$ given by Williams [7] and van den Berg and Lewis [3]. In local coordinates the corresponding equations have a particularly simple form. Let E, F and G be the fundamental magnitudes of the first order defined by

$$E = \left|\frac{\partial X}{\partial u_1}(u)\right|^2 , \quad F = \frac{\partial X}{\partial u_1}(u) \cdot \frac{\partial X}{\partial u_2}(u) , \quad G = \left|\frac{\partial X}{\partial u_2}(u)\right|^2$$

then

$$du_1(t) = \frac{1}{2}\Delta u_1 \, dt - \frac{1}{(EG-F^2)^{1/2}} \frac{\partial X}{\partial u_2}(u) \cdot dB(t) \qquad (2.20)$$

$$du_2(t) = \frac{1}{2}\Delta u_2 \, dt + \frac{1}{(EG-F^2)^{1/2}} \frac{\partial X}{\partial u_1}(u) \cdot dB(t) \qquad (2.21)$$

We feel that these equations should be given more prominence as they are so useful for computer simulations. We should emphasise that these equations are valid for coordinates u_1 and u_2 which are not necessarily orthogonal. Using these equations in the next section we consider some examples of Brownian motion on parametrized surfaces embedded in $\mathbb{R}^3$.

3. Examples

(i) Surface of Revolution

Let $V : I \times \mathbb{R} \to \mathbb{R}^3$ be the parametrized surface of revolution obtained by rotating the parametrized curve, $c(s) = (\alpha(s), \beta(s))$ $\beta(s) > 0$ $\forall s \in I$, about the α-axis. Thus

$$V(s,\theta) = \{\underset{\sim}{x} \in \mathbb{R}^3 : \underset{\sim}{x}(s,\theta) = (\alpha(s), \beta(s)\cos\theta, \beta(s)\sin\theta) \; s \in I, \theta \in \mathbb{R}\} . \qquad (3.1)$$

For a surface of this kind, parametrized in s and θ, the operator Δ is given by, suppressing the dependence of α and β on s,

$$\Delta = \frac{1}{\beta(\alpha'^2+\beta'^2)^{1/2}} \left\{ \frac{\partial}{\partial s} \left(\frac{\beta}{(\alpha'^2+\beta'^2)^{1/2}} \frac{\partial}{\partial s} \right) + \frac{(\alpha'^2+\beta'^2)^{1/2}}{\beta} \frac{\partial^2}{\partial \theta^2} \right\} . \qquad (3.2)$$

Using equations (2.20) and (2.21) of the previous section and the above, a Brownian motion on V, starting at $\underset{\sim}{x}_o = \underset{\sim}{x}_o(s_o, \theta_o)$, is characterized by the following stochastic differentials of the parametric coordinates s and θ :-

$$ds(t) = \frac{1}{2(\alpha'^2+\beta'^2)} \left\{ \frac{\beta'}{\beta} - \frac{\alpha'\alpha''+\beta'\beta''}{(\alpha'^2+\beta'^2)} \right\} dt + \frac{1}{(\alpha'^2+\beta'^2)^{1/2}} (dB_t^2 \sin\theta - dB_t^3 \cos\theta) \qquad (3.3)$$

$$d\theta(t) = \frac{1}{\beta(\alpha'^2+\beta'^2)^{1/2}} \{\alpha' dB_t^1 + \beta'(dB_t^2 \cos\theta + dB_t^3 \sin\theta)\} , \qquad (3.4)$$

where $\underset{\sim}{B}(t) = (B_t^1, B_t^2, B_t^3)$ is the driving Brownian motion in $\mathbb{R}^3$.

(ii) Pseudosphere

If we take

$$c(s) = \left(\int_0^s (1 - e^{-2u/c})^{1/2}\, du\ ,\ c\, e^{-s/c}\right),\quad s \in [0,\infty),\quad c > 0\ , \tag{3.5}$$

in the above example, the surface V is the parametrized pseudosphere in $\mathbb{R}^3$. i.e. V has constant negative Gaussian curvature $K = -1/c^2$. (Here s is the arc length along the surface.)

The stochastic differentials for s and θ characterizing a Brownian motion on V are, by (3.3) and (3.4),

$$ds(t) = -\frac{1}{2c}dt + dB_t^2 \sin\theta - dB_t^3 \cos\theta \tag{3.6}$$

$$d\theta(t) = \frac{1}{c}\{(e^{2s/c} - 1)^{1/2}\, dB_t^1 - dB_t^2 \cos\theta - dB_t^3 \sin\theta\}\ . \tag{3.7}$$

(iii) Torus

Let $V : \mathbb{R}^2 \to \mathbb{R}^3$ be the parametrized surface of revolution obtained by rotating the parametrized circle $c(\phi) = (a + b(1+\cos\phi), b\sin\phi)$ $(a,b > 0)$, in the (x,z) plane, about the z-axis. Thus

$$V(\theta,\phi) = \{\underset{\sim}{x} \in \mathbb{R}^3 : \underset{\sim}{x}(\theta,\phi) = ([a+b(1+\cos\phi)]\cos\theta, [a+b(1+\cos\phi)]\sin\theta, b\sin\phi) : (\theta,\phi) \in \mathbb{R}^2\} \tag{3.8}$$

is the parametrized torus in $\mathbb{R}^3$.

For the torus parametrized in θ and ϕ the operator Δ is given by

$$\Delta = \frac{1}{[a+b(1+\cos\phi)]}\left\{\frac{1}{[a+b(1+\cos\phi)]}\frac{\partial^2}{\partial\theta^2} + \frac{1}{b^2}\frac{\partial}{\partial\phi}([a+b(1+\cos\phi)]\frac{\partial}{\partial\phi})\right\}\ . \tag{3.9}$$

Hence the stochastic differentials for θ and ϕ characterizing a Brownian motion on V, starting at $\underset{\sim}{x}_o = \underset{\sim}{x}_o(\theta_o,\phi_o)$, are given by

$$d\theta(t) = \frac{1}{[a+b(1+\cos\phi)]}\{\sin\phi\,(dB_t^1 \cos\theta + dB_t^2 \sin\theta) - dB_t^3 \cos\phi\} \tag{3.10}$$

$$d\phi(t) = \frac{-\sin\phi}{2b[a+b(1+\cos\phi)]}dt + \frac{1}{b}(dB_t^2 \cos\theta - dB_t^1 \sin\theta)\ . \tag{3.11}$$

In the next section we discuss computer simulations of these motions.

4. Computer Simulation

Note. The computer simulations herein were performed on the B.B.C. model B microcomputer. We shall, therefore, occasionally refer to the use of its basic commands. For a fuller description of these commands see the User Guide.

With our characterization of Brownian motion on surfaces, computer simulations become very simple and accurate to perform, as no approximations are necessary in order to keep the motion on the surface. The only difficulty lies in drawing the surface in 3-D perspective view, with the appropriate hidden detail. The method we use, although not a general hidden surface algorithm, is a version of the 'back to front' method. With this method it is possible to draw many simple surfaces, including those considered in section 3, quite quickly and with the appropriate hidden detail.

Using a discrete parametrization (usually orthogonal) the surface is approximated by dividing it into a number of quadrilateral shaped facets. The surface is constructed by plotting and colouring these facets in a particular sequence. Using the PLOT85 and GCOLØ commands (to plot and colour each facet) the areas furthest from the eye are drawn first and the nearest last. In this way the further unseen areas are overdrawn, thus producing the necessary hidden detail. Being able to draw a surface, we now need to generate the driving Brownian motion increments in $\mathbb{R}^3$.

Using the random number generator on $(0,1)$ we can generate four independent uniformly distributed random variables, $U_1 = U(0,1)$, $U_2 = U(0,2\pi)$, $U_3 = U(0,1)$, and $U_4 = U(0,2\pi)$. Then via the transformations,

$$N_1 = \surd(-2\,dt\,\log U_1)\cos U_2\,, \quad N_2 = \surd(-2\,dt\,\log U_1)\sin U_2\,,$$
$$N_3 = \surd(-2\,dt\,\log U_3)\cos U_4\,, \quad (\text{or } N_3 = \surd(-2\,dt\,\log U_3)\sin U_4)\,. \tag{4.1}$$

(N_1,N_2,N_3) are three independent normally distributed random variables with mean zero and variance $\surd dt$. i.e. (N_1,N_2,N_3) are the required Brownian increments.

The following listing, which should clarify most of the above ideas, is of a program which simulates Brownian motion on a torus. The figs (4.1) and (4.2) show computer simulations of Brownian motion on a pseudosphere and torus respectively. Also shown in figs (4.3) and (4.4) are computer simulations of Brownian motion on a Möbius band and sphere. We leave these as exercises.

```
 10 MODE1
 20 PROCDETAIL
 30 PROCPOINTS
 40 PROCTORUS
 50 PROCMOTION
 60 END
 70 DEF FNX(K,L,M)
 80  XN = R%*E%*O%*K
 90  XD = D% + O%*(NP*L - NN*M)
100  = XN/XD
110 DEF FNY(K,L,M)
120  YN = E%*O%*(NN*L + NP*M)
130  = YN/XD
```

```
140 DEF PROCMOTION
150  GCOL0,1
160  T = RAD(-90) : P = RAD(90)
170  H = A% + B%*(1 + COSP)
180  X = H*COST : Y = H*SINT : Z = B%*SINP
190  MOVE FNX(X,Y,Z),FNY(X,Y,Z)
200  U1 = RND(1) : U2 = 2*PI*RND(1)
210  U3 = RND(1) : U4 = 2*PI*RND(1)
220  R1 = SQR(-2*dt*LNU1) : R2 = SQR(-2*dt*LNU3)
230  B1 = R1*COSU2 : B2 = R1*SINU2 : B3 = R2*COSU4
240  DT = ((B1*COST + B2*SINT)*SINP - B3*COSP)/H
250  DP = C*(-dt*SINP/(2*H) + B2*COST - B1*SINT)
260  T = T + DT : P = P + DP
270  H = A% + B%*(1 + COSP)
280  X = H*COST : Y = H*SINT : Z = B%*SINP
290  DRAW FNX(X,Y,Z),FNY(X,Y,Z)
300  GOTO 200
310 ENDPROC
320 DEF PROCTORUS
330  GCOL0,131 : CLG
340  FOR N% = 0 TO 3
350  IF N% = 0 THEN Q% = 19 : F% = 0 : S% = -1
360  IF N% = 1 THEN Q% = 10 : F% = 9 : S% = 1
370  IF N% = 2 THEN Q% = 0 : F% = 0 : S% = 1
380  IF N% = 3 THEN Q% = 9 : F% = 9 : S% = -1
390  FOR J% = Q% TO Q% + 9*S% STEP S%
400  FOR I% = F% TO F% + 8
410  MOVE X%(I%+1,J%),Y%(I%+1,J%)
420  MOVE X%(I%,J%),Y%(I%,J%)
430  GCOL0,0
440  PLOT85, X%(I%+1,J%+1),Y%(I%+1,J%+1)
450  PLOT85, X%(I%,J%+1),Y%(I%,J%+1)
460  GCOL0,2
470  DRAW X%(I%,J%),Y%(I%,J%)
480  DRAW X%(I%+1,J%),Y%(I%+1,J%)
490  DRAW X%(I%+1,J%+1),Y%(I%+1,J%+1)
500  DRAW X%(I%,J%+1),Y%(I%,J%+1)
510  NEXT I%
520  NEXT J%
530  NEXT N%
540 ENDPROC
550 DEF PROCPOINTS
560  FOR I% = 0 TO 18
570  T% = 20*I% : T = RADT%
580  FOR J% = 0 TO 20
590  P% = 18*J% : P = RADP%
600  H = A% + B%*(1 + COSP)
610  X = H*COST : Y = H*SINT : Z = B%*SINP
620  X%(I%,J%) = INT (FNX(X,Y,Z) + 0.5)
630  Y%(I%,J%) = INT (FNY(X,Y,Z) + 0.5)
640  NEXT J%
650  NEXT I%
660 ENDPROC
670 DEF PROCDETAIL
680  DIM X%(18,20) : DIM Y%(18,20)
690  VDU29,639;511; : VDU19,3,4,0,0,0
700  TH = RAD(40) : R% = 700 : O% = 10
710  E% = 10000 - R% : D% = R%*(E% + R%)
720  NP = R%*COSTH : NN = R%*SINTH
730  A% = 1 : B% = 30 : C = 1/B% : dt = 1
740 ENDPROC
```

Fig. (4.1)

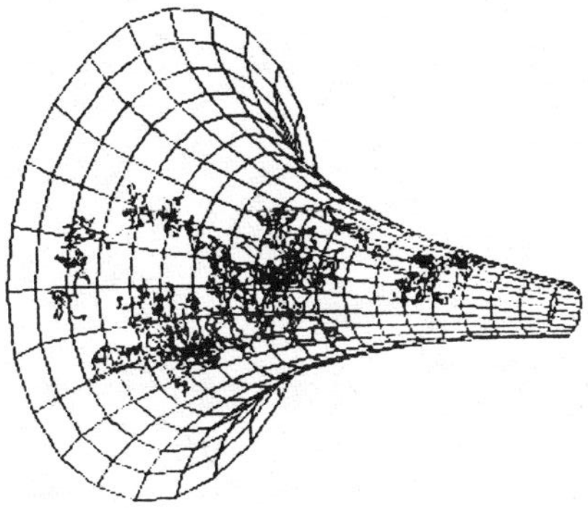

Fig. (4.2)

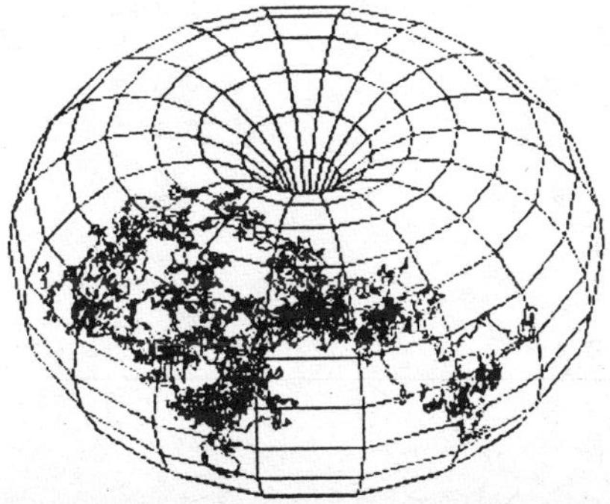

Fig. (4.3)

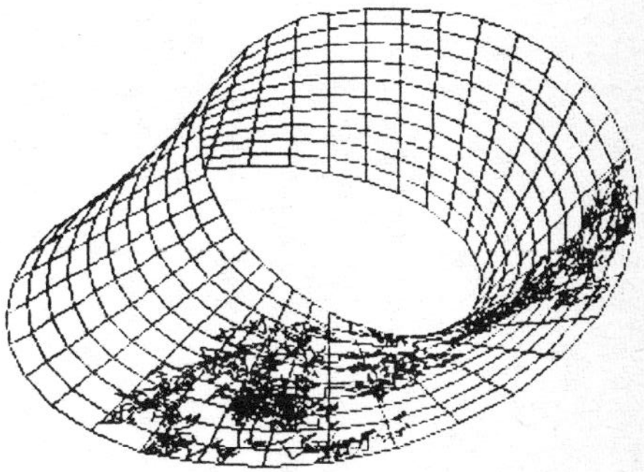

Fig. (4.4)

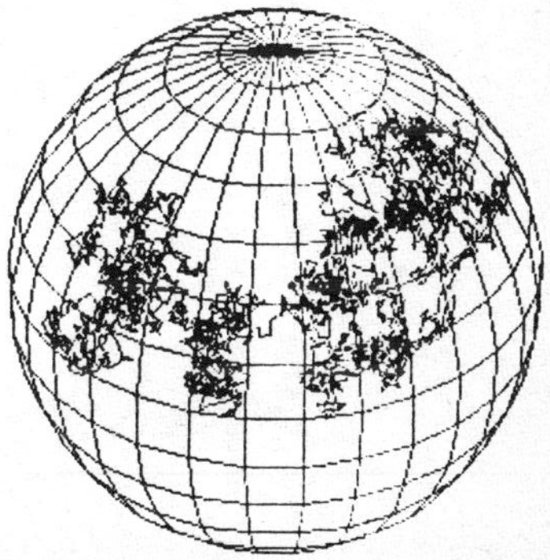

5. Computer Simulations of Stochastic Processes of the Hydrogen Atom

The principal aim behind these simulations is to produce, via stochastic mechanics, a pictorial representation of the charge distributions in the excited states of hydrogen. The simulations which we performed are very simple.

Using the stochastic process for the state ψ_{klm} (usual notation) arising from Nelson's stochastic mechanics, we ran a single sample path, only recording its position on the screen when it entered a thin cross section through the origin and centred in a plane of constant azimuth ϕ. This position was recorded in a special way. Utilizing the POINT command, which tests the colour at any point on the screen, we colour-coded the number of impacts the process made at any particular point, thus building up a colour-coded picture of the cumulative frequency of impacts corresponding to the charge distribution. For states with nodes separate processes were run in each non-communicating nodal region.

The following pictures show shaded print outs for the states ψ_{200}, ψ_{210}, and ψ_{211}, and also for the particularly interesting ψ_{321} state. The solid black curves represent the classical polar diagrams scaled by the factor, $\langle r \rangle_\psi$, the expected value of the radial coordinate. The arrows indicate the direction of the z-axis and the governing stochastic differential equation is given in spherical polar coordinates. Scales are given in terms of a_o, the Bohr radius.

Immediately below in Figure 1 we see the "exponential convergence" of sample paths for different starting points for the ground state process of the hydrogen atom. In the paper by Elworthy and Chappell in this same volume the corresponding Lyapunov exponent is shown to be negative definite.

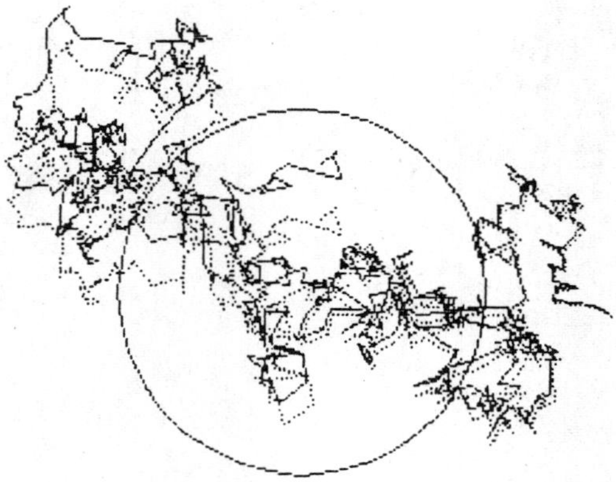

Figure 1

State ψ_{200}

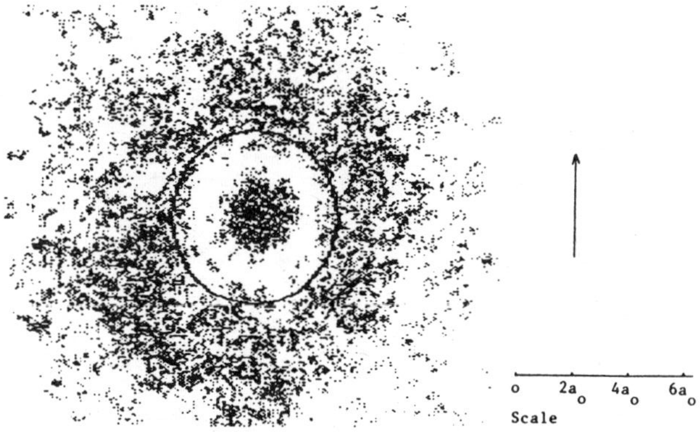

$$d\underset{\sim}{X}(t) = -\frac{\hbar}{m}\left\{\frac{1}{2a_o - r} + \frac{1}{2a_o}\right\}\hat{\underset{\sim}{r}}\,dt + \left(\frac{\hbar}{m}\right)^{1/2} d\underset{\sim}{B}(t)\,.$$

State ψ_{211}

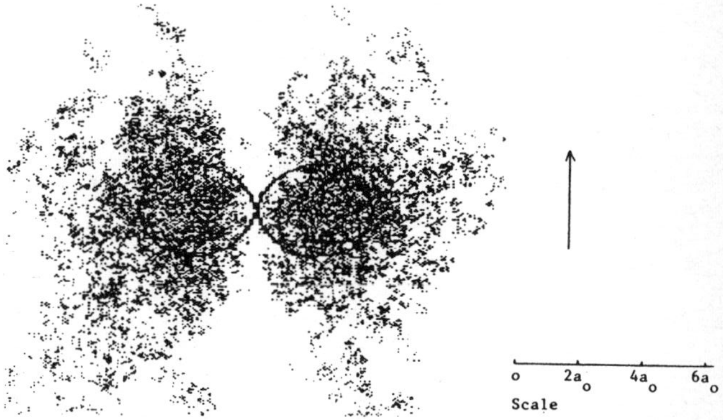

$$d\underset{\sim}{X}(t) = -\frac{\hbar}{m}\left\{\left(\frac{1}{2a_o} - \frac{1}{r}\right)\hat{\underset{\sim}{r}} - \frac{\cot\theta}{r}\hat{\underset{\sim}{\theta}} - \frac{\operatorname{cosec}\theta}{r}\hat{\underset{\sim}{\phi}}\right\}dt + \left(\frac{\hbar}{m}\right)^{1/2} d\underset{\sim}{B}(t)\,.$$

State ψ_{210}

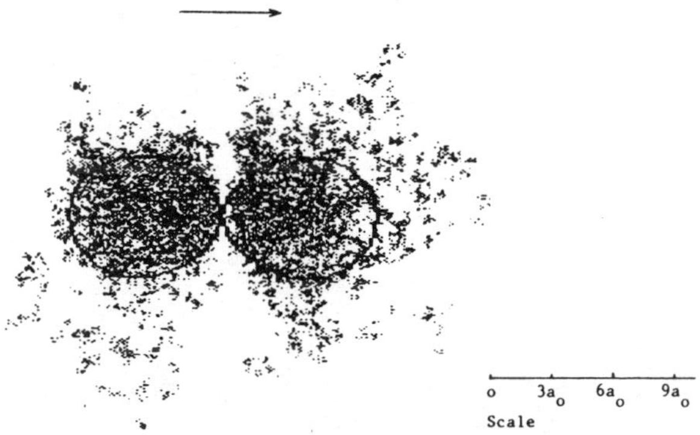

$$d\underset{\sim}{X}(t) = -\frac{\hbar}{m}\left\{\left(\frac{1}{2a_o} - \frac{1}{r}\right)\hat{\underset{\sim}{r}} + \frac{\tan\theta}{r}\hat{\underset{\sim}{\theta}}\right\}dt + \left(\frac{\hbar}{m}\right)^{1/2} d\underset{\sim}{B}(t)\,.$$

State ψ_{321}

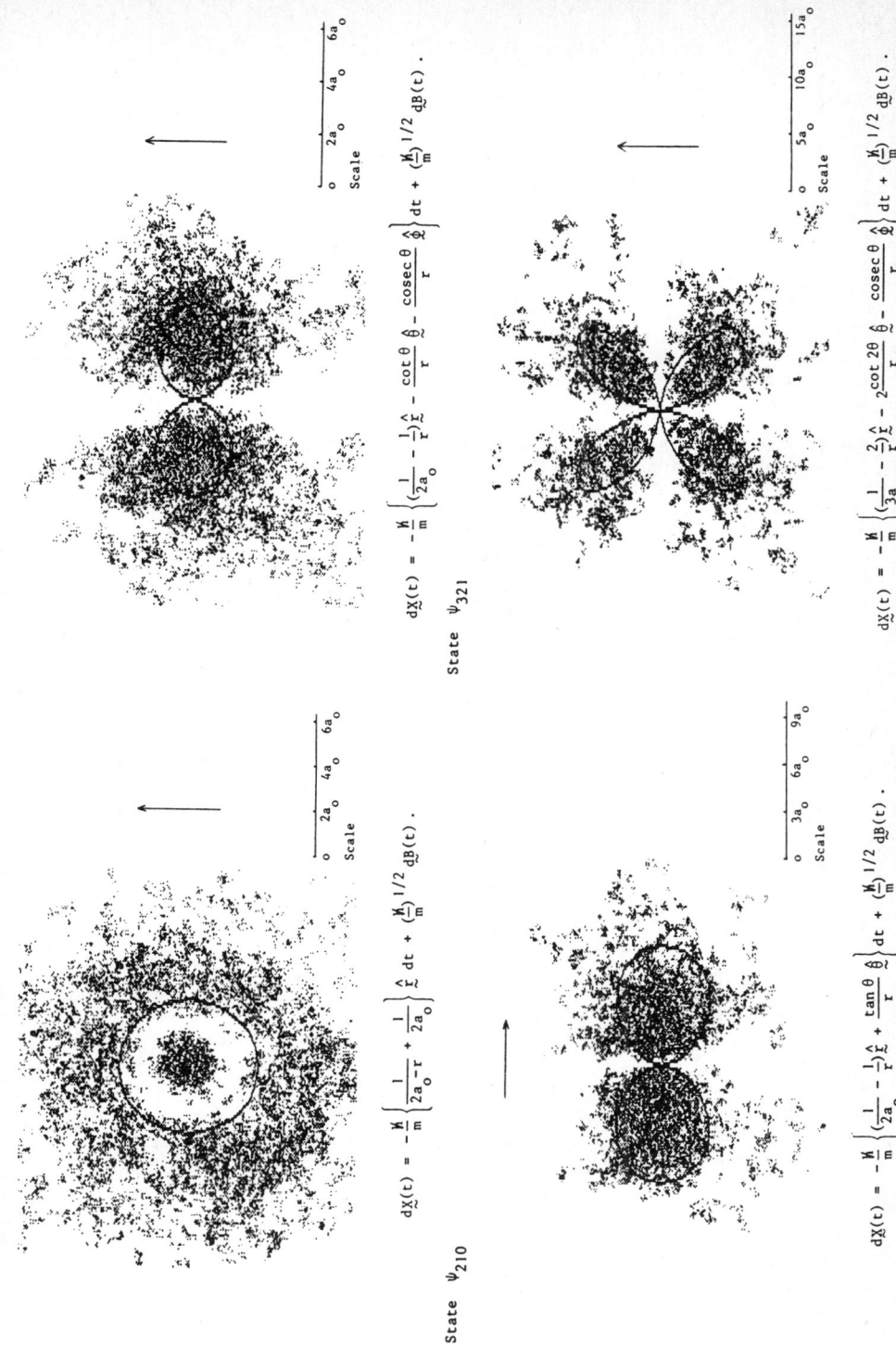

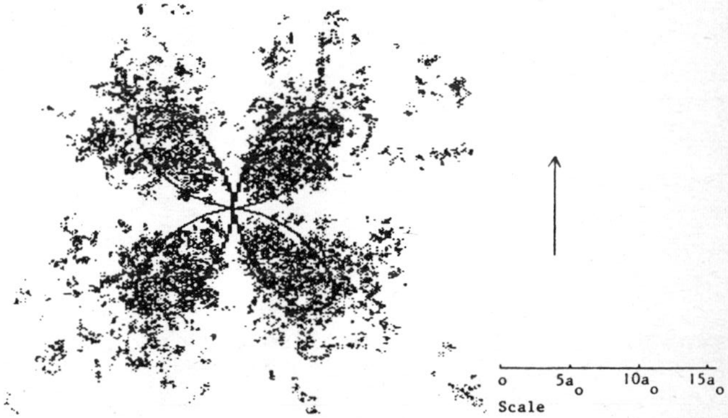

$$d\underset{\sim}{X}(t) = -\frac{\hbar}{m}\left\{\left(\frac{1}{3a_o} - \frac{2}{r}\right)\hat{\underset{\sim}{r}} - 2\frac{\cot 2\theta}{r}\hat{\underset{\sim}{\theta}} - \frac{\operatorname{cosec}\theta}{r}\hat{\underset{\sim}{\phi}}\right\}dt + \left(\frac{\hbar}{m}\right)^{1/2} d\underset{\sim}{B}(t)\,.$$

Acknowledgement

It is a pleasure to thank Professor David Elworthy for helpful conversations concerning some of this work.

References

[1] I.O. Angell and B.J. Jones. Advanced Graphics with the BBC Model B Microcomputer, Macmillan, 1983.

[2] P.H. Baxendale. Wiener Processes on Manifolds of Maps, Proc. Royal Soc. Edinburgh 87A (1980), 127-152.

[3] M. van den Berg and J.T. Lewis. Brownian Motion on a Hypersurface, Bull. London Math. Soc. 17 (1985), 144-150.

[4] R.W.R. Darling. A Martingale on the Embedded Torus, Bull. London Math. Soc. 15 (1985), 221-225.

[5] K.D. Elworthy. Stochastic Differential Equations on Manifolds, Cambridge University Press, 1982.

[6] J.T. Lewis. An Elementary Approach to Brownian Motion on Manifolds. In 'Stochastic Processes - Mathematics and Physics', Proceedings, Bielefeld 1984. Lecture Notes in Mathematics No. 1158, Springer-Verlag, 1985.

[7] G.C. Price and D. Williams. Rolling with "Slipping": I, Séminaire de Probabilitiés XVII, 1981/82, Paris. Lecture Notes in Mathematics No. 986, Springer-Verlag, 1983.

[8] J.A. Thorpe. Elementary Topics in Differential Geometry, Undergraduate Texts in Mathematics, Springer-Verlag, 1979.

[9] C.E. Weatherburn. Differential Geometry of Three Dimensions, Vol. 1, Cambridge University Press, 1931.

[10] H.E. White. Pictorial Representations of the Electron cloud for Hydrogen-like Atoms, Phys. Rev. 37 (1931), 1416-1424.

SUPERSYMMETRY AND STOCHASTIC PROCESSES

Z. Haba
Institute of Theoretical Physics
University of Wrocław, Poland

Stochastic description of quantum mechanics enables a treatment of quantum phenomena in a classical way. It provides a better understanding of classical aspects of the quantum mechanics. In many cases the probabilistic approach leads to solutions of quantum problems in a more direct and simple way (see refs. [1]-[4]). In application to elementary particle physics the stochastic approach requires an extension at least in two directions i) infinite dimensional stochastic processes or random fields ii) a description of fermions. Stochastic processes in infinite dimensional spaces with application to (Euclidean) quantum field theory have been discussed in refs. [5]-[6]. The stochastic equations, which one gets this way, can be considered as ordinary equations in an infinite dimensional space or as parabolic partial differential equations for random fields. We shall discuss here another approach based on elliptic equations for random fields, which are explicitly Euclidean invariant. In relation to the stochastic mechanics (see Carlen's paper [7] on stochastic field theory) it must be emphasized that we restrict ourselves to the ground state process i.e. to the Euclidean quantum field theory (see ref. [8]).

If we follow the conventional quantum field theory of Fermi fields, then the quantum Fermi field has as a classical limit an anticommuting object i.e. an element of the Grassmann algebra. So, we need a probability theory on the Grassmann algebra in order to stochastically quantize anticommuting fields. It may be possible to describe anticommutativity in the framework of classical probability i.e. solely in terms of commuting random variables. However, the progress of supersymmetry (a symmetry between commuting and anticommuting variables) suggests that such an approach may be a futile one from the point of view of applications.

A notion of a non-commutative stochastic process has been discussed before by many authors (see refs. [9]-[12]). We think that the anticommutative case should have a much simpler form closer to the classical probability theory. As an example, the differential geometry has a direct generalization to the anticommutative variables [13] , which is much simpler (and richer), than the general non-commutative theory [14].

In this paper we shall define an anticommuting Brownian motion and construct its realization in the Fock space (sec.I), then we discuss stochastic equations and a connection between the Brownian motion and differential equations (sec.II). Finally in sec. III we consider applications to supersymmetric models currently studied in particle physics.

I. Anticommuting Brownian motion

Let Λ be a finite dimensional Grassmann algebra over real numbers and M an infinite dimensional one ($\Lambda \subset M$). We define a real anticommuting Brownian motion $\xi_\Theta = \Theta + \xi$ starting from $\Theta \in \Lambda$ by means of the formula for expectation values

$$E[\xi^\alpha(t)\ \xi^\beta(s)] = C^{\alpha\beta}\min(t,s) \tag{1}$$

$$E[\xi_{\alpha_1}(t_1)\ \ldots\ldots\ \xi_{\alpha_{2n}}(t_{2n})] = \det[C^{\alpha_i\alpha_j}\ \min(t_i,t_j)] \tag{2}$$

the expectation of an odd number of ξ's vanishing. In eqs. (1)-(2) E[ ] is a linear functional on the Grassmann algebra M, which can have various realizations. C is a real invertible antisymmetric matrix.

We could consider as well the anticommuting white noise

$$a^\alpha(t) = \frac{d}{dt}\ \xi^\alpha(t) \tag{3}$$

$$E[a^\alpha(t)\ a^\beta(s)] = C^{\alpha\beta}\delta(t-s) \tag{4}$$

as our primary object, then $\xi^\alpha(t) = \int_0^t a^\alpha(\tau)d\tau$.

Assume that C has the form

$$C = \begin{pmatrix} \sigma & & 0 \\ & \ddots & \\ 0 & & \sigma \end{pmatrix} \qquad \text{where} \qquad \sigma = \begin{pmatrix} 0 & 1 \\ -1 & 0 \end{pmatrix} = (\varepsilon^{\alpha\beta})$$

Then, we can define a complex Brownian motion. If C is a 2×2 matrix, then

$$\hat{\xi} = 1/\sqrt{2}\ (\xi^2 + \xi^1) \quad , \qquad \xi = 1/\sqrt{2}\ (\xi^2 - \xi^1)$$

$$E[\hat{\xi}(t)\ \xi(s)] = \min(t,s) \tag{5}$$

Next, if $C = \begin{pmatrix} \sigma & 0 \\ 0 & \sigma \end{pmatrix}$ is a 4×4 matrix, $(\xi) = (\lambda, \eta)$

$$\upsilon = 1/\sqrt{2}\ (\hat{\lambda} + i\eta) \quad , \qquad \hat{\upsilon} = 1/\sqrt{2}(\hat{\lambda} - i\eta)$$

$$\hat{\upsilon}^* = 1/\sqrt{2}\ (\lambda + i\hat{\eta}) \quad , \qquad \upsilon^* = 1/\sqrt{2}(\lambda - i\hat{\eta}) \tag{6}$$

Then

$$E[\upsilon(t)\ \hat{\upsilon}^*(s)] = E[\hat{\upsilon}(t)\ \upsilon^*(s)] = \min(t,s) \tag{7}$$

the remaining expectations vanishing.
Next, using the complex Brownian motion (6) we can define

$$\hat{\psi}(t) = \frac{d}{dt}\ \hat{\upsilon}(t) \qquad\qquad \psi^*(t) = \upsilon^*(t)$$

In such a case from eq.(7) we get

$$E[\hat{\psi}(t)\ \psi^*(s)] = \Theta(s-t) = (-\frac{d}{d\tau})^{-1}(t,s) \tag{8}$$

There can exist various realizations of the linear functional E[ ] defined on the Grassmann algebra M by eqs.(1)-(2). An analogue of the Wiener measure called the Berezin integral does not have a sound mathematical basis, yet. So, we construct first a well-defined realization of the anticommuting Brownian motion in the Fock space. This realization may be applied for a rigorous theory of the differential equations of sec.II. However, the Fock space does not seem to be a proper realization if we wish to construct stationary anticommuting processes discussed in sec.III. For this reason we shall discuss the Berezin integral at the end of this section.

Let H be a Hilbert space. Then, the fermionic Fock space F is just a direct sum of antisymmetric tensor products $\Lambda^n H$ of H

$$F = \text{Ext}\, H = \oplus H_n$$

where $H_o = R$ and $H_n = \Lambda^n H$. The element $1 \in H_o$ is called the Fock vacuum. The scalar product

$$(f_1 \wedge \dots \wedge f_n,\ g_1 \wedge \dots \wedge g_n) = \delta_{m,n}\ \det[(f_i, g_j)] \tag{9}$$

where (,) is the scalar product in H, determines the Hilbert space structure in F. Define the creation operator $A^+(f)$ $(f \in H)$

$$A^+(f)(f_1 \wedge \dots \wedge f_n) = f \wedge f_1 \wedge \dots \wedge f_n \tag{10}$$

The annihilation operator A(f) is the Hermitian adjoint with respect to the scalar product (9). In such a case

$$\{A^+(f), A(g)\} = A^+(f)A(g) + A(g)A^+(f) = (f,g)$$

In algebra the operator A(g) is denoted by i(g) and called interior multiplication (while $A^+(f)$ is the exterior multiplication according to eq.(10)).
If we choose a basis Θ_a in H then

$$A_a^+ = \Theta_a \wedge \qquad \text{and} \qquad A_a = \frac{\partial}{\partial \Theta_a} \tag{11}$$

where $\frac{\partial}{\partial \Theta^a}$ is an antiderivation on the Grassmann algebra.

We choose $H = L^2(R) \times V$, where V is 2n-dimensional real vector space (if C is $2n \times 2n$ matrix). Then, $A^+(f)$ is a linear functional on H

$$A^+(f) = \int A_\alpha^+(t) f^\alpha(t) dt \qquad \text{where} \quad \{A_\alpha^+(t), A_\beta(s)\} = \delta_{\alpha\beta}\delta(t-s) \tag{12}$$

Now, assuming (for simplicity) that C consists of σ-matrices on the diagonal, we can express each real white noise (4) in terms of the creation and annihilation operators

$$a_{\alpha i}(t) = A_{\alpha i}(t) - \varepsilon_{\alpha\beta} A_{\beta i}^+(t) \tag{13}$$

$\alpha=1,2$, $i=1,2,\ldots,n$ (then $\xi(t)=\int_0^t a(\tau)d\tau$). In such a case the expectation E[ ] is an expectation value in the Fock vacuum 1 (we have $A \cdot 1=0$)

$$E[\Phi(\xi)] = (1, \Phi(A,A^+)1) \tag{14}$$

It is easy to check that with the definitions (13)-(14) equations (1)-(8) are satisfied.

Fermionic creation and annihilation operators are bounded operators in F (bounded operators will be denoted by $B(F)$ from now on). In fact

$$|A(f)| = |A^+(f)| = (f,f)^{\frac{1}{2}}$$

Hence, if we choose an orthonormal basis $\{f_b\}$ $(b=(n,\alpha))$ in H and define

$$a_N(f) = \sum_{n \leq N} a(f_b)(f_b, f)$$

where $a(f_b)=a_\alpha(n)$ fulfil

$$E[a_\alpha(n) a_\beta(m)] = \varepsilon_{\alpha\beta}\delta_{nm} \tag{15}$$

then clearly $|a(f) - a_N(f)| \to 0$.

So, we could define the Brownian motion ξ_t as a limit in the operator norm of independent Gaussian random variables $a_\alpha(n)$.

For a finite number of variables the expectation E[ ] has a realization in terms of the Berezin integral. This is a linear functional on the Grassman algebra determined by the formulas

$$\int a_\beta(m)da_\alpha(n) = \delta_{\alpha\beta} \qquad \int 1\cdot da_\alpha(m) = 0$$

Then

$$E[\Phi] = \int \prod_{n,\alpha} da_\alpha(n)\exp\left(\sum_n^N \varepsilon^{\alpha\beta}a_\alpha(n)a_\beta(n)\right)\Phi(a) \tag{16}$$

The integral (16) has a nice interpretation as an analogue of the Gaussian integral on Ext$H$ [15]-[16]. For a finite number of variables the Fock (14) and the Berezin (16) realizations coincide. We can control the limit $N\to\infty$ of the operator $\Phi(\xi_N)$ on the l.h.s. of eq.(16) through the Fock space realization (14). Unfortunately, there does not exist an integration theory on Ext$H$, when H is infinite dimensional. Even though the expectation values have a limit when $N\to\infty$, it is not evident how to define the limit of anticommuting variables in terms of the Berezin integration, as the Berezin integral does not determine a norm on the Grassmann algebra.

II. Differential equations

We can follow now the usual procedure to define the stochastic integral of $f\in B(F)$ (stochastic integrals and the Ito formula for non-commutative processes have been studied before in refs. [10]-[11])

$$\int_0^t f d\xi = \lim_{N\to\infty} \sum f(\xi(\tau_i))(\xi(\tau_{i+1})-\xi(\tau_i)) \qquad \text{(Ito)}$$

$$\int_0^t f\circ d\xi = \lim_{N\to\infty} \sum f(\xi(\tfrac{1}{2}(\tau_{i+1} + \tau_i)))(\xi(\tau_{i+1})-\xi(\tau_i)) \qquad \text{(Stratonovitch)} \tag{17}$$

where $\tau_i = it/N$ and the limit is understood in the operator norm of $B(F)$. In order to show the existence of the limit (17) it is sufficient to compute $\| I_{N,M}\psi\|^2 = ((I_N-I_M)\psi, (I_N-I_M)\psi)$, where I_N denotes the sum in (17) with f being a polynomial in ξ and $\psi\in\Lambda^n H$. In such a case, the expression for $\| I_{N,M}\psi\|$ has the same form (up to the $C^{\alpha\beta}$ metric) as for the standard Wiener process. Hence, $\| I_{N,M}\psi\|\to 0$.

In a similar way we can establish the Ito formula: if $f\in B(F)$ is three times Frechet differentiable (in the operator norm) then

$$f(\xi(t)) = f(\xi(0)) + \int_0^t \partial_\alpha f(\xi(\tau))\circ d\xi^\alpha(\tau) =$$
$$= f(\xi(0)) + \int_0^t \partial_\alpha f(\xi(\tau))d\xi^\alpha(\tau) + \tfrac{1}{2}C^{\alpha\beta}\int_0^t \partial_\alpha\partial_\beta f(\xi(\tau))d\tau \tag{18}$$

where $\partial_\alpha f$ is defined by the formula

$$\Theta^\alpha\partial_\alpha f(\xi) = \lim_{\rho\to 0} 1/\rho\ (f(\xi + \rho\Theta) - f(\xi))$$

According to eq.(11)

$$\partial_\alpha f(\xi) = \frac{\partial}{\partial\xi^\alpha} f = i(\alpha)f$$

∂_α is the algebraic antiderivation $i(\alpha)$ determined by $i(\alpha)\xi^\beta=\delta^{\alpha\beta}$ and the Leibnitz rule.

Consider next the stochastic equation

$$d\zeta = \beta(\zeta)dt + \gamma(\zeta)d\xi \tag{19}$$

In order to relate solutions of stochastic equations to partial differential equations we need a notion of conditional expectation and a Markov property. Let $F_t = =\{\xi(\tau)\in M, \tau\leqq t\}$ be the σ-algebra (a subalgebra of the Grassmann algebra M) generated by $\xi(\tau)$ $(\tau\leqq t)$, where the closure is defined with respect to the operator norm in $\mathcal{B}(F)$. Then, we define the conditional expectation $\mathcal{E}$ with respect to F_t as a linear operator $\mathcal{E}_s \in L(\mathcal{B}(F),\mathcal{B}(F))$ defined for $f\in\mathcal{B}(F)$ by

$$\mathcal{E}_s f = (E[f])^{-1}E_s[f] = (E[f])^{-1}(P_s f + E[(1-P_s)f]) \tag{20}$$

where P_s is the projector onto F_s i.e. the expectation value (1) is computed only over ξ_τ with $\tau>s$ e.g.

$$E_s[(\int_o^t f(\xi)d\xi)^2] = (\int_o^s f(\xi)d\xi)^2 + \int_o^s f(\xi)d\xi E[\int_s^t f(\xi)d\xi]$$

$$+ E[\int_s^t f(\xi)d\xi]\int_o^s f(\xi)d\xi + E[(\int_s^t f(\xi)d\xi)^2]$$

$\mathcal{E}_t$ fulfills the Markov property

$$\mathcal{E}_t\, \mathcal{E}_s = \mathcal{E}_t \qquad \text{if} \qquad t\leqq s$$

Moreover, if $\sigma_t F_s = F_{s+t}$ is the time translation, then $\sigma_t \mathcal{E}_s = \mathcal{E}_{s+t}\sigma_t$ as a consequence of the invariance of the Brownian motion (1) under the time shift. Hence, if we define $T_t = \mathcal{E}_o\sigma_t$, then for $f\in F_o$

$$T_t T_s f = \mathcal{E}_o\sigma_t\, \mathcal{E}_o\sigma_s f = \mathcal{E}_o\mathcal{E}_t\sigma_{s+t} f = \mathcal{E}_o\sigma_{s+t} f = T_{s+t} f$$

It follows in the usual way that a non-anticipating solution of the stochastic equation (19) defines a semigroup on F_o

$$(T_t f)(\Theta) = E_o[f(\Theta+\zeta_t)] \tag{21}$$

where $\zeta_o=0$ and $\Theta\in\Lambda$.

There are several problems concerning the formulas (20)-(21), which we did not study, yet; i) Is E_s a Hermitian projector in a certain Hilbert space? ii) Is the semigroup T_s (21) Hermitian? The first question is of mathematical relevance concerning an interpretation in the framework of non-commutative probability theory in the sense of Segal [9]. The second problem is of physical relevance, because we wish the generator of T_s (the Hamiltonian) to be self-adjoint in the physical Hilbert space. In the commutative case the generator of the stochastic process is self-adjoint in $L^2(d\mu)$, where μ is the invariant measure. We shall discuss the invariant measure later on.

Let us compute the generator $H=\frac{dT}{dt}\Big|_{t=0}$ of the semigroup (21). Under the assumption that f is analytic we get

$$H = \frac{1}{2}\gamma^{\alpha}{}_{\rho}(\Theta)C^{\rho\nu}\gamma^{\mu}{}_{\nu}(\Theta)\frac{\partial}{\partial\Theta^{\alpha}}\frac{\partial}{\partial\Theta^{\mu}} + \beta^{\mu}(\Theta)\frac{\partial}{\partial\Theta^{\mu}} \tag{22}$$

We can define the transition function $p_t(\Theta,\Theta')$ by means of the formula

$$(T_t f)(\Theta) = \int \prod_{\alpha} d\Theta'_{\alpha} p_t(\Theta,\Theta')f(\Theta') \tag{23}$$

where $d\Theta$ is the Berezin integral (16) (well-defined here, because the number of Θ's is finite) e.g. if

$$H = \frac{1}{2}C^{\alpha\beta}\frac{\partial}{\partial\Theta^{\alpha}}\frac{\partial}{\partial\Theta^{\beta}}$$

then

$$p_t(\Theta,\Theta') = (\det C)^{-\frac{1}{2}}(2t)^n \exp\left(-1/_{2t}\,(\Theta_{\alpha}-\Theta'_{\alpha})C^{\alpha\beta}(\Theta_{\beta}-\Theta'_{\beta})\right) \tag{24}$$

p_t fulfills the diffusion equation

$$\partial_t p_t = H p_t \tag{25}$$

with the initial condition $p_o(\Theta,\Theta') = \delta(\Theta,\Theta')$, where the δ-function is defined by

$$f(\Theta) = \int \prod_{\alpha} d\Theta'_{\alpha}\delta(\Theta,\Theta')f(\Theta') \tag{26}$$

It can be checked that

$$\delta(\Theta,\Theta') = \prod_{\alpha}(\Theta_{\alpha}-\Theta'_{\alpha})$$

Assume now that $\gamma=1$ in eq.(19) and $C^2=-1$. Then, the semigroup (21) can be expressed in the form (the Girsanov formula)

$$(T_t f)(\Theta) = E_o[\exp(\int_o^t \beta^\alpha C_{\alpha\rho} d\xi^\rho + \tfrac{1}{2}\int_o^t \beta^\alpha C_{\alpha\rho}\beta^\rho) f(\Theta + \xi_t)] \tag{27}$$

Eq.(27) can be proved through differentiation using the Ito formula (17).

Let us consider a special case of eq.(19) (the fermionic oscillator) $\gamma=1$, $\beta(\zeta) = -\Omega\zeta$. Then, the stochastic equation

$$d\zeta = -\Omega\zeta dt + d\xi$$

has the solution

$$\zeta_t = e^{-\Omega t}\zeta_o + \int_o^t e^{-\Omega(t-\tau)} d\xi(\tau) \tag{28}$$

There exists a stationary solution, if the operator Ω is positive in L $(\mathcal{B}(F), \mathcal{B}(F))$ (which is a Hilbert space with the trace as the scalar product). Then,

$$\zeta_t = \int_{-\infty}^t e^{-\Omega(t-\tau)} d\xi(\tau) \tag{29}$$

In the linear case the stochastic integral in eq.(27) can be evaluated by means of the Ito formula (17). We get

$$(T_t f)(\Theta) = E_o[\exp\tfrac{1}{2}(\Omega\xi(t)C\xi(t) - \Omega\Theta C\Theta - t\mathrm{Tr}\Omega)\exp(\tfrac{1}{2}\int_o^t \Omega\xi C\Omega\xi d\tau) f(\Theta + \xi_t)] \tag{30}$$

Let us define the invariant measure μ for the stochastic process ζ (19) as the Berezin measure such that

$$\int d\mu(\Theta) f(\Theta) = \int d\mu(\Theta) \prod_\alpha d\Theta'_\alpha p_t(\Theta,\Theta') f(\Theta') \tag{31}$$

It can be checked that

$$d\mu(\Theta) = \prod_\alpha d\Theta_\alpha (\det\tfrac{1}{2}\Omega)^{-\frac{1}{2}} \exp(\tfrac{1}{2}\Omega\Theta C\Theta) \tag{32}$$

is the invariant measure for the fermionic oscillator. Moreover, H is Hermitian in $L^2(d\mu)$ in the sense

$$\int d\mu f\cdot Hg = \int d\mu\, Hf\cdot g \tag{33}$$

Next, it follows from eq.(16) that $L^2(d\mu)$ has a realization as a fermionic Fock space. Hence, H is Hermitian in this Fock space.

Eq.(33), i.e. a formal Hermicity of H (22) in $L^2(d\mu)$, can be proved in general through direct calculations. From eq.(31) it follows that $H^*\mu=0$. Then, the Hermicity can be established through integration by parts in eq.(33). However, in general the scalar product in $L^2(d\mu)$ is not positive definite.

III. Models

We shall discuss now an application of the anticommuting Brownian motion to some models of particle physics involving Fermi fields. The models are supersymmetric i.e. there is a symmetry between commuting and anticommuting variables. We are working in the framework of Euclidean quantum field theory. The continuation to the real time can be achieved through the continuation of exp-tH to expitH.

A. Brownian motion on a supergroup [17]

Let $\Lambda=\Lambda_o\oplus\Lambda_1$, where Λ_o is the commuting part of the Grassmann algebra and Λ_1 is the anticommuting one. Then, $A^{m,n}=\Lambda_o^m\times\Lambda_1^n$ is called the superspace. The supermanifold is a Hausdorf topological space, which is locally diffeomorphic to $A^{m,n}$. Then, the supergroup is defined as a supermanifold with a group composition law defined on it. In such a case the Lie algebra of left invariant vector fields consists of derivations as well as antiderivations.

We define the Brownian motion

$$\xi^A(t) = (b^i(t),\xi^\alpha(t)) \tag{34}$$

$i=1,\ldots,m,\quad \alpha=1,\ldots,2n$
on the superspace A, where b^i is the standard Wiener process and ξ^α is the anticommuting Brownian motion (1) i.e.

$$E[\xi^A(t)\ \xi^B(s)] = h^{AB}\min(t,s) \tag{35}$$

where $h^{AB} = (\delta^{ij},C^{\alpha\beta})$.

Let us restrict ourselves here to matrix realizations of the supergroup G. Let τ_A be a matrix realization of the algebra of G. Then, we define the Brownian motion on the supergroup G as the solution of the stochastic equation ($M\in G$)

$$dM = M\circ d\xi \tag{36}$$

where $d\xi = d\xi^A\tau_A$

Eq. (36) is a linear equation, hence it is explicitly soluble in terms of time ordered exponentials.

As a special case of eq.(36) let us consider the Brownian motion on the Euclidean super Poincaré group $\mathcal{P}$. The algebra consists of the commutators between the generators of rotations M_{ij} and translations P_k and the anticommutators

$$\{Q_\alpha, Q_\beta\} = -(\gamma_k C)_{\alpha\beta} P_k \tag{37}$$

where γ's form a Majorana realization of the Clifford algebra, C is the charge conjugation matrix. The restriction of the Brownian motion (36) to the superspace $\mathcal{M} = \mathcal{P}/O(n)$ has the generator

$$H = \tfrac{1}{2}\partial_k\partial_k + \tfrac{1}{2}l^{-1}\bar{D}D$$

where $\bar{D}^\alpha = C^{\alpha\beta}D_\beta$ and

$$D_\beta = \frac{\partial}{\partial\Theta^\beta} - i/2(\gamma_k\Theta)_\beta\partial_k$$

Such a Hamiltonian for a spinning particle has been derived in another way in refs. [18].

B. Supersymmetric quantum mechanics

This is a quantum theory of a classical model with commuting as well as anticommuting variables. The model has been introduced and studied by Witten [19]. In order to describe the model in terms of stochastic equations let us introduce the superfield depending on two additional independent Grassmann variables Θ and $\hat{\Theta}$

$$B(t,\Theta,\hat{\Theta}) = b(t) + \hat{\Theta}\psi(t) + \Theta\hat{\psi}(t) + i\Theta\hat{\Theta}\frac{d\hat{b}}{dt} \tag{39}$$

In this formula ψ and $\hat{\psi}$ are the anticommuting processes defined in eq.(8), while b and $\hat{b}$ are independent Wiener processes. In terms of the superfield the supersymmetry transformation takes the simple form

$$B(t,\Theta,\hat{\Theta}) \to B(t + \varepsilon\Theta + \hat{\varepsilon}\hat{\Theta}, \Theta+\varepsilon, \hat{\Theta}+\hat{\varepsilon})$$

where ε and $\hat{\varepsilon}$ are Grassmann variables. The left invariant vector field corresponding to the ε-transformation has the form

$$D = \frac{\partial}{\partial\Theta} + \Theta\frac{\partial}{\partial t}$$

We restrict ourselves here to the supersymmetric particle moving on a matrix group G Let τ_a be a matrix realization of the algebra of G and $B=B^a\tau_a$. Then, the stochastic equation for a superfield $g(t,\Theta,\hat{\Theta})\in G$ reads

$$Dg = g \circ DB \tag{40}$$

Eq. (40) is linear, hence explicitly soluble. The equation (as well as the solution) on a homogeneous space G/H can be derived from eq. (40) by means of a projection $G \to G/H$ (see [20]-[21]). The Hamiltonian is defined as usual as a generator of the stochastic process. It can be shown that $H = dd^* + d^* d$ (the Laplace operator on differential forms). It appears that the anticommuting Brownian motion plays in this model only an auxiliary role as a description of differential forms. In fact, the semigroup generated by H could be expressed in the form $(T_t f)(\omega) = E[f(\Phi_t^* \omega)]$, where Φ_t^* denotes the pull-back of the flow Φ_t determined by the standard Brownian motion on G/H and acting on the Maurer-Cartan differential forms ω over G/H.

C. Supersymmetric field theory

It appears that complex fields and complex manifolds play a distinguished role in a generalization of stochastic equations to two-dimensional field theories (see ref. [21]). The two-dimensional complex massless free scalar field defined as the Gaussian field with the covariance

$$E[\bar{\phi}(x)\ \phi(y)] = (-\Delta)^{-1}(x,y) \tag{41}$$

is a counterpart to the Brownian motion, while

$\eta = \dfrac{\partial \phi}{\partial \bar{z}}$ is the noise

$$E\left[\overline{\frac{\partial \phi}{\partial \bar{z}}}\ \frac{\partial \phi}{\partial \bar{z}'}\right] = \delta^{(2)}(z - z') \tag{42}$$

here $\dfrac{\partial}{\partial \bar{z}} = \dfrac{\partial}{\partial x^0} + i\dfrac{\partial}{\partial x^1}$.

Consider as an example fields with values in a sphere S^2 parametrized by the stereographic projection onto the complex plane C (coordinates $w \in C$). Then, there is a natural analogue of the conventional stochastic equation

$$\frac{\partial w}{\partial \bar{z}} = (1 + \bar{w}w)\ \frac{\partial \phi}{\partial \bar{z}} \tag{43}$$

If w does not depend on x_1 and $\dfrac{\partial \phi}{\partial \bar{z}}$ is the complex white noise (independent of x_1), then the stochastic equation (43) is reduced to the conventional (Ito) equation for the Brownian motion on the sphere.

We wish to generalize eq. (43) to anticommuting variables. Let us introduce the supernoise $\eta(x,\Theta,\hat{\Theta})$ (DB in eq. (40) is a 1-dimensional realization of the supernoise)

$$E[\eta^+(z,\Theta,\hat{\Theta})\ \eta(z',\Theta',\hat{\Theta}')] = \delta^{(2)}(z-z')\delta(\Theta-\Theta')\delta(\hat{\Theta}-\hat{\Theta}') \tag{44}$$

where $\delta(\Theta,\Theta')$ has been defined in eq. (26).

Let D be the supersymmetric derivative

$$D = \frac{\partial}{\partial\Theta} + \Theta \frac{\partial}{\partial\bar{z}}$$

Then, we have shown in ref. [22] that the equation

$$DW = (1 + W^{+}W)\eta \quad (45)$$

describes supersymmetric fields with values in a sphere. W is a superfield built from scalar as well as spinor fields. The Hamiltonian corresponding to the solution of eq. (45) acts in a space $L^2(d\mu)$, where $d\mu$ is an infinite dimensional Berezin integral. The construction of a solution of eq. (45) poses a difficult mathematical problem.

REFERENCES

1 G. Jona-Lasinio, F. Martinelli and E. Scopolla, Comm.Math.Phys.80,223(1981)
2 S. Albeverio, R. Hoegh-Krohn and L.Streit, J.Math.Phys.18,907(1977)
3 K.D. Elworthy and A. Truman, J.Math.Phys.22,2144(1981)
4 B.Simon, Functional Integration and Quantum Physics, Academic, 1979
5 S. Albeverio and R. Hoegh-Krohn, Z. Wahr. v. Geb.40,1(1977)
6 G. Jona-Lasinio and P.K. Mitter, Comm.Math.Phys.101,409(1985)
7 E. Carlen, these Proceedings
8 F. Guerra and P. Ruggiero, Phys.Rev.Lett.31,1022(1973)
9 I.E. Segal, Ann.Math.57,401(1953)
10 C.Barnett, R.F. Streater and I. Wilde, Journ.Funct.Anal.48,172(1982)
11 D.B. Applebaum and R. Hudson, Comm.Math.Phys.96,473(1984)
12 L. Accardi, A. Frigerio and J.T. Lewis, Publ.RIMS 18,97(1982)
13 B. de Witt, Supermanifolds, Cambridge Univ., 1984
14 A. Connes, IHES Publ., 1986
15 L. Gross, in "Les Meth. Math.Th. Quan des Champs", CNRS Paris, 1976
16 V. Mathai and D. Quillen, Topol.25,85(1986)
17 Z. Haba, BiBOS - Bielefeld preprint, No 214
18 R. Casalbuoni, Nuovo Cimento 33A,389(1976)
L. Brink and J.H. Schwarz, Phys.Lett.100B,310(1981)
D.V. Volkov and A.Y. Pashnev, Theor.Math.Phys.44,321(1980)
19 E. Witten, Nucl.Phys.B202,253(1982)
J.Diff.Geom.17,661(1982)
20 Z. Haba, in Lect.Notes in Phys., No 262, S. Albeverio, G. Casati and D. Merlini Eds., Springer 1986
21 Z. Haba, Phys.Rev.D33,2428(1986)
22 Z. Haba, Phys.Rev.D in print

Note added: We have learned that just recently Alice Rogers discussed a realization by means of the Berezin integral of an analogue of the Brownian motion (8) in the preprint "Fermionic path integration and Grassmann Brownian motion" (King's College, December 1986)

ALGEBRAIC THEORY OF QUANTUM DIFFUSIONS

R L Hudson
Department of Mathematics
University of Nottingham
Nottingham NG72RD
England

Abstract
Quantum diffusions are quantum stochastic processes in the sense of Accardi, Frigerio and Lewis, in which evolution of elements of the initial algebra is governed by a system of autonomous quantum stochastic differential equations against the gauge, creation and annihilation processes. As a consequence of the quantum Itô formula the coefficients of these equations satisfy cohomological identities. A diffusion for which the coefficients differ in a cohomologically trivial sense from a given diffusion can be constructed by a perturbation procedure. Every quantum diffusion on the algebra of all bounded operators on a Hilbert space is characterised by a unitary process. In the commutative case certain "diffusions" with discontinuous sample paths are found; these are a feature of the zero temperature Fock quantum stochastic calculus used here and do not exist at finite temperature.

§1. Introduction

In the formulation of Accardi, Frigerio and Lewis [1], a quantum stochastic process is characterised as a quadruple $(\mathcal{A},\tilde{\mathcal{A}},\tilde{\omega},j)$ comprising unital $\star$-algebras $\mathcal{A}$ and $\tilde{\mathcal{A}}$, a state $\tilde{\omega}$ of $\tilde{\mathcal{A}}$, and a family $j = (j_t : t \in \mathbb{R}_{\geq 0})$ of injective homomorphisms from $\mathcal{A}$ into $\tilde{\mathcal{A}}$.

In this paper we shall consider processes of this type, in which $\mathcal{A}$ is an algebra of bounded operators on a Hilbert space $\mathfrak{H}_0$, and $\tilde{\mathcal{A}}$ is the algebra $B(\tilde{\mathfrak{H}})$ of all bounded operators on the Hilbert space $\tilde{\mathfrak{H}} = \mathfrak{H}_0 \otimes \mathfrak{H}$, where $\mathfrak{H}$ is the Boson Fock space over $L^2(\mathbb{R}_{\geq 0})$. j_0 will always be the ampliation $x \mapsto x \otimes I$. The state $\tilde{\omega}$ will be the tensor product of a state ω_0 of $\mathcal{A}$ with the vacuum state on $B(\mathfrak{H})$.

Following [4], we define the <u>gauge</u>, <u>creation</u> and <u>annihilation</u> processes in $\mathfrak{H}$ as the families of operators defined on the algebraic tensor product of $\mathfrak{H}_0$ with the span of the exponential vectors (coherent states) $\psi(f)$, $f \in L^2(\mathbb{R}_{\geq 0})$ in $\mathfrak{H}$ by the actions

$$\left.\begin{aligned} \Lambda(t)u \otimes \psi(f) &= \frac{d}{d\varepsilon} u \otimes \psi\left(\exp \varepsilon\chi_{[0,t]} f\right)\Big|_{\varepsilon=0} \\ A^{\dagger}(t)u \otimes \psi(f) &= \frac{d}{d\varepsilon} u \otimes \psi(f+\varepsilon\chi_{[0,t]})\Big|_{\varepsilon=0} \\ A(t)u \otimes \psi(f) &= \int_{[0,t]} f \; u \otimes \psi(f) \end{aligned}\right\} u \in \mathfrak{H}_0, f \in L^2(\mathbb{R}_{\geq 0}).$$

These should not be envisaged as quantum stochastic processes in the sense of [1] but, in the present context, regarded as mathematical devices with the aid of which quantum stochastic processes are constructed, just as classical Brownian motion is used to construct more physically realistic models, even of physical Brownian motion.

The mechanism of the construction is the quantum stochastic calculus of [4] which gives meaning to stochastic integrals of certain operator-valued processes against the integrators $d\Lambda$, $dA^{\dagger}$ and dA. Thus we assume that, for each $x \in \mathcal{a}$, $x_t := j_t(x_o)$ constitutes an adapted process in the sense of [4], and that these processes satisfy quantum stochastic differential equations of the form

$$dx_t = j_t(\lambda(x_o))d\Lambda + j_t(\alpha(x_o))dA^{\dagger} + j_t(\alpha^{\dagger}(x_o))dA + j_t(\tau(x_o))dt. \quad (1.1)$$

Here λ, α, $\alpha^{\dagger}$ and τ are maps frm $\mathcal{a}$ to itself; we call them the <u>structure maps</u> of the <u>quantum diffusion</u> j on $\mathcal{a}$. We write (1.1) succinctly as

$$dx = \lambda(x)d\Lambda + \alpha(x)dA^{\dagger} + \alpha^{\dagger}(x)dA + \tau(x)dt,$$

but emphasise that the structure maps are from $\mathcal{a}$ to itself and that for instance $\lambda(x)$ is a notational abbreviation for the adapted process $j_t(x_o))$. Thus quantum diffusions are governed by systems of autonomous quantum stochastic differential equations.

The fact that each j_t is a homomorphism enforces properties of the structure maps. They inherit linearity from the linearity of j_t. Since j_t is a $*$-map, $\lambda = \lambda^{\dagger}$, $\tau = \tau^{\dagger}$ and $\alpha^{\dagger}$ is indeed the adjoint map of α, defined by

$$\alpha^{\dagger}(x) = (\alpha(x^*))^*.$$

Since $j_t(I) \equiv I$ the structure maps vanish on I.

To exploit the multiplicativity of j_t we require the technical condition that j_t is contractive for the uniform operator norm. This is automatic if $\mathcal{a}$ is a C^*-algebra or if it is closed under formation of holomorphic functions of self-adjoint elements (for instance if $\mathcal{a}$ is the algebra $C^{\infty}(V)$ of smooth functions on a compact Riemann manifold V acting on $L^2(V)$ by multiplication).

When each j_t is contractive, for arbitrary x_o and y_o in $\mathcal{a}$ the stochastic differentials dx and dy satisfy the condition ensuring validity of the quantum Ito product formula of [4]. Comparing coefficients of the basic differ ntials in

$$d(xy) = dx\, y + x\, dy + dx\, dy$$

further identities to be satisfied by the structure maps are obtained; these are called the <u>restraint equations</u>.

This calculation is carried out in §2. In §3 it is shown that the restraint equations are cohomological in character and the classification of their solutions is considered. In §4 the effect of replacing the structure maps by cohomologically equivalent maps is examined , and it is shown that the new diffusion can be constructed from the old by a perturbation procedure. In §5 examples are considered; in particular when $\mathcal{Q} = B(\mathfrak{h}_0)$ every quantum diffusion on $\mathcal{Q}$ is shown to be completely described by a unitary process of the type constructed in[4]. In contrast, when $\mathcal{Q}$ is commutative, as well as classical diffusions, a class of quantum diffusions related to generalised Poisson processes is constructed which are not governed by such unitary processes.

I wish to thank several colleagues and students, particularly David Applebaum, Frank Ball, Mark Evans and Michael Schürmann for valuable discussions and comments on a first draft of this work.

§2. Restraint equations

Recall from [4], Theorem 4.5, that if

$$dM_k = E_k d\Lambda + F_k dA^+ + G_k dA + H_k dt, \quad k=1,2$$

are stochastic differentials satisfying

$$\sup_{0\leq s\leq t}\max_k\left\{||M_k(s)||,||E_k(s)||,||F_k(s)||,||G_k(s)||,||H_k(s)||\right\}<\infty \tag{2.1}$$

for each $t>0$, then $d(M_1M_2)$ is also a stochastic differential, given by

$$d(M_1M_2) = M_1 dM_2 + dM_1\, M_2 + dM_1\, dM_2$$

where the basic differentials $d\Lambda$, dA^+, dA and dt commute with adapted processes and the Itô correction dM_1dM_2 is evaluated by combining this with extension by bilinearity of the basic multiplication rules

	$d\Lambda$	dA^+	dA	dt
$d\Lambda$	$d\Lambda$	dA^+	0	0
dA^+	0	0	0	0
dA	dA	dt	0	0
dt	0	0	0	0

(2.2)

Now let j be a quantum diffusion on $\mathcal{Q}$. Provided each j_t is contractive, for arbitrary $x_0, y_0 \in \mathcal{A}$, the stochastic differentials

$$dx = \lambda(x)d\Lambda + \alpha(x)dA^{\dagger} + \alpha^{\dagger}(x)dA + \tau(x)dt$$
$$dy = \lambda(y)d\Lambda + \beta(y)dA^{\dagger} + \alpha^{\dagger}(y)dA + \tau(y)dt$$

satisfy the condition (2.1), so that we may write

$$d(xy) = x\,dy + dx\,y + dx\,dy. \tag{2.3}$$

On the other hand

$$d(xy) = \lambda(xy)d\Lambda + \alpha(xy)dA^{\dagger} + \alpha^{\dagger}(xy)dA + \tau(xy)dt. \tag{2.4}$$

Using (2.2) to compare coefficients of the basic differentials between (2.3) and (2.4) we obtain the <u>restraint equations</u>

$$\lambda(xy) = x\lambda(y) + \lambda(x)y + \lambda(x)\lambda(y) \tag{2.5}$$
$$\alpha(xy) = x\alpha(y) + \alpha(x)y + \lambda(x)\alpha(y) \tag{2.6}$$
$$\tau(xy) = x\tau(y) + \tau(x)y + \alpha^{\dagger}(x)\alpha(y). \tag{2.7}$$

These may be regarded as identities to be satisfied by the structure maps for arbitrary $x, y \in \mathcal{A}$.

§3. Solutions of the restraint equations

By adding xy to both sides of (2.5), we see that $\lambda + \mathrm{id}$ is multiplicative on $\mathcal{A}$, where id is the identity map. Since λ is a $*$-map and $\lambda(I) = 0$ we may therefore set

$$\lambda = \sigma - \mathrm{id} \tag{3.1}$$

where σ is a unital $*$-endomorphism of $\mathcal{A}$. Conversely, given any such σ, λ defined by (3.1) is a $*$-map vanishing on I which satisfies (2.5).

Substituting (3.1) into (2.6) gives

$$\alpha(xy) = \sigma(x)\alpha(y) + \alpha(x)y \tag{3.2}$$

for all $x, y \in \mathcal{A}$, that is α must be a <u>σ-derivation</u> of $\mathcal{A}$. For fixed $\ell \in \mathcal{A}$

$$\alpha(x) \equiv \ell x - \sigma(x)\ell$$

defines a σ-derivation which we denote by $\mathrm{ad}_{\sigma}\ell$; such σ-derivations are called <u>inner</u>.

<u>Proposition</u> 3.1 Suppose that σ ia an inner automorphism of $\mathfrak{G}$, that is, there exists a unitary $w\in\mathfrak{G}$ such that $\sigma(x) \equiv wxw^{-1}$. Suppose further that every derivation of $\mathfrak{G}$ is inner. Then every σ-derivation of $\mathfrak{G}$ is inner.

<u>Proof</u> Let α be a σ-derivation, so that for arbitrary $x,y\in\mathfrak{G}$,

$$\begin{aligned}\alpha(xy) &= \sigma(x)\alpha(y) + \alpha(x)y\\ &= wxw^{-1}\alpha(y) + \alpha(x)y.\end{aligned}$$

Multiplying both sides by w^{-1} shows that $\tilde{\alpha}(x) \equiv w^{-1}\alpha(x)$ is a derivation. By hypothesis $\tilde{\alpha} = \mathrm{ad}\,\tilde{\ell}$ for some $\tilde{\ell}\in\mathfrak{G}$. But then, for all $x\in\mathfrak{G}$,

$$\begin{aligned}\alpha(x) &= w\tilde{\alpha}(x)\\ &= w\tilde{\ell}x - wx\tilde{\ell}\\ &= \ell x - \sigma(x)\ell\end{aligned}$$

where $\ell = w\tilde{\ell}$. Thus $\alpha = \mathrm{ad}_\sigma\ell$. □

The third restraint equation (2.7) may be rewritten as

$$\delta\tau(xy) = -\alpha^{+}(x)\alpha(y) \tag{3.3}$$

where, for linear $\tau:\mathfrak{G} \quad \mathfrak{G}$, $\delta\tau:\mathfrak{G}\times\mathfrak{G}\mapsto\mathfrak{G}$ is defined by

$$\delta\tau(x,y) = x\tau(y) - \tau(xy) + \tau(x)y. \tag{3.4}$$

In the language of the Hochschild algebra cohomology theory in which $\mathfrak{G}$ is its own two-sided module via left and right multiplication, the 2-cochain η defined by

$$\eta(x,y) = -\alpha^{+}(x)\alpha(y) \tag{3.5}$$

must therefore be a 2-coboundary. The next proposition shows that, in this language, η is at least a 2-cocycle.

<u>Proposition</u> 3.2 Let α be a σ-derivation. Then, if η is defined by (3.5), $\delta\eta \equiv 0$, where $\delta\eta:\mathfrak{G}\times\mathfrak{G}\times\mathfrak{G}\mapsto\mathfrak{G}$ is defined by

$$\delta\eta(x,y.z) = x\eta(y,z) - \eta(xy,z) + \eta(x,yz) - \eta(x,y)z.$$

Proof The adjoint map α^{+} of the σ-derivation α satisfies the identity

$$\alpha^{\dagger}(xy) = \alpha^{\dagger}(x)\sigma(y) + x\alpha^{\dagger}(y).$$

Hence for arbitrary $x,y,z\in\mathfrak{a}$

$$\begin{aligned}\delta\eta(x,y,z) &= -x\alpha^{\dagger}(y)\alpha(z) + \alpha^{\dagger}(xy)\alpha(z) - \alpha^{\dagger}(x)\alpha(yz) + \alpha^{\dagger}(x)\alpha(y)z \\ &= -x\alpha^{\dagger}(y)\alpha(z) + \alpha^{\dagger}(x)\sigma(y)\alpha(z) + x\alpha^{\dagger}(y)\alpha(z) \\ &\quad -\alpha^{\dagger}(x)\sigma(y)\alpha(z) - \alpha^{\dagger}(x)\alpha(y)z + \alpha^{\dagger}(x)\alpha(y)z = 0. \square\end{aligned}$$

If the second order cohomology space is nontrivial, there may exist σ-derivations for which the cocycle η is not a coboundary, so that there exists no $\tau\colon \mathfrak{a} \mapsto \mathfrak{a}$ satisfying (3.3). An example is when $\alpha = \alpha_1 + i\alpha_2$ where α_1 and α_2 are the basic outer derivations of the algebra of "smooth elements of the noncommutative torus" [7,5]. The next proposition shows that this circumstance can only arise when α is not inner. Thus the restraint equations offer topological obstructions to the construction of quantum diffusions only when both first and second order cohomology spaces are nontrivial.

Proposition 3.3 Let $\alpha = \mathrm{ad}_{\sigma}\ell$ be an inner σ-derivation. Then if

$$\tau(x) \equiv -\tfrac{1}{2}(\ell^{*}\ell x - 2\ell^{*}\sigma(x)\ell + x\ell^{*}\ell), \tag{3.6}$$

τ satisfies (2.7).

Proof With τ given by (3.6), from (3.4)

$$\begin{aligned}\delta\tau(x,y) &= -\tfrac{1}{2}[x\ell^{*}\ell y - 2x\ell^{*}\sigma(y)\ell + xy\ell^{*}\ell - \ell^{*}\ell xy \\ &\quad + 2\ell^{*}\sigma(xy)\ell - xy\ell^{*}\ell + \ell^{*}\ell xy - 2\ell^{*}\sigma(x)\ell y + x\ell^{*}\ell y] \\ &= -[x\ell^{*}\ell y + \ell^{*}\sigma(x)\ (y)\ell - x\ell^{*}\sigma(y)\ell - \ell^{*}\sigma(x)\ell y] \\ &= -(x\ell^{*} - \ell^{*}\sigma(x))(\ell y - \sigma(y)\ell) \\ &= -\alpha^{\dagger}(x)\alpha(y).\end{aligned}$$

Thus (3.3), which is equivalent to (2.7), is satisfied. $\square$

Proposition 3.4 Suppose that $\lambda = 0$, so that $\sigma = \mathrm{id}$, and that $\alpha = \alpha^{\dagger}$ is a self-adjoint derivation. Then if τ is defined by

$$\tau(x) = \tfrac{1}{2}\alpha(\alpha(x)),$$

τ satisfies (2.7).

Proof Direct verification. □

If a particular solution τ of (2.7) has been found for a given σ-derivation α, for instance by applying Propositions 3.3 or 3.4, the general solution is got by adding an arbitrary self-adjoint derivation to this particular solution.

A triple (λ,α,τ) of linear maps from $\mathcal{A}$ to itself satisfying the conditions necessary for them to define the coefficients of the quantum stochastic differential equations of a quantum diffusion, that is, all vanishing on I, such that λ and τ are $*$-maps and satisfying the restraint equations (2.5), (2.6) and (2.7), is called a restraint triple.

§4. Perturbations

Theorem 4.1 Let (λ,α,τ) be a restraint triple and set $\sigma = \lambda + \mathrm{id}$. Let $w_o, \ell_o, h_o \in \mathcal{A}$ with w_o unitary and h_o self-adjoint. Define maps $\tilde{\lambda}=\tilde{\sigma} - \mathrm{id}$, $\tilde{\alpha}$ and $\tilde{\tau}$ from $\mathcal{A}$ to $\mathcal{A}$ by

$$\tilde{\sigma}(x) = w_o\sigma(x)w_o^* \tag{4.1}$$

$$\tilde{\alpha}(x) = \ell_o x - w_o\sigma(x)w_o^*\ell_o + w_o\alpha(x) \tag{4.2}$$

$$\begin{aligned}\tilde{\tau}(x) &= \tau(x) + i\,\mathrm{ad}h_o(x) - \tfrac{1}{2}(\ell_o^*\ell_o x - 2\ell_o^* w_o\sigma(x)w_o^*\ell_o + x\ell_o^*\ell_o) \\ &\quad -\ell_o^* w_o\alpha(x) - \alpha^+(x)w_o^*\ell_o.\end{aligned} \tag{4.3}$$

Then $(\tilde{\lambda},\tilde{\alpha},\tilde{\tau})$ is a restraint triple.

Proof $\tilde{\sigma}$ is clearly a unital endomorphism of $\mathcal{A}$ so $\tilde{\lambda}$ satisfies (2.5). Since α is a σ-derivation, for arbitrary $x,y\in\mathcal{A}$,

$$\begin{aligned}w_o\alpha(xy) &= w_o\sigma(x)\alpha(y) + w_o\alpha(x)y \\ &= \tilde{\sigma}(x)w_o\alpha(y) + w_o\alpha(x)y,\end{aligned}$$

showing that $x \longmapsto w_o\alpha(x)$ is a $\tilde{\sigma}$-derivation. Thus $\tilde{\alpha}$ is the sum of this $\tilde{\sigma}$-derivation and the inner $\tilde{\sigma}$-derivation $\mathrm{ad}_{\tilde{\sigma}}\ell_o$ and is itself a $\tilde{\sigma}$-derivation as required.

$\tilde{\tau}$ is manifestly a $*$-map vanishing on the identity. To establish (2.7), that is that

$$\delta\tilde{\tau}(x,y) = -\tilde{\alpha}^+(x)\tilde{\alpha}(y) \tag{4.4}$$

for all $x,y\in\mathcal{A}$, we write

$$\tilde{\tau} = \tau_o + \tau_1 + \tau + \rho \tag{4.5}$$

where, for $x \in \mathcal{A}$,

$$\tau_0(x) = i\,\mathrm{ad}h_0(x),$$
$$\tau_1(x) = -\tfrac{1}{2}(\ell_0^*\ell_0 x - 2\ell_0^*\tilde{\sigma}(x)\ell_0 + x\ell_0^*\ell_0),$$
$$\rho(x) = -\ell_0^* w_0 \alpha(x) - \alpha^\dagger(x) w_0^* \ell_0,$$

and, correspondingly,

$$-\tilde{\alpha}^\dagger(x)\tilde{\alpha}(y) = T_1(x,y) + T(x,y) + R(x,y)$$

where

$$T_1(x,y) = -(\mathrm{ad}_{\tilde{\sigma}}\ell_0)^\dagger(x)(\mathrm{ad}_{\tilde{\sigma}}\ell_0)(y),$$
$$T(x,y) = -\alpha^\dagger(x)\alpha(y)$$
$$R(x,y) = \alpha^\dagger(x)w_0^*(\mathrm{ad}_{\tilde{\sigma}}\ell_0)(y) - (\mathrm{ad}_{\tilde{\sigma}}\ell_0)^\dagger(x)w_0\alpha(y).$$

Applying the coboundary operation δ defined by (3.4) to (4.5), we have that $\delta\tau_0 = 0$ since τ_0 is a derivation, $\delta\tau_1 = T_1$ by Proposition 3.3 and $\delta\tau = T$ by hypothesis. Thus it remains to be shown that $\delta\rho = R$. For $x, y \in \mathcal{A}$,

$$\begin{aligned}\delta\rho(x,y) &= -x\ell_0^* w_0\alpha(y) - x\alpha^\dagger(y)w_0^*\ell_0 + \ell_0^* w_0\alpha(xy)\\ &\quad +\alpha^\dagger(xy)w_0^*\ell_0 - \ell_0^* w_0\alpha(x)y - \alpha^\dagger(x)w_0^*\ell_0 y\\ &= -x\ell_0^* w_0\alpha(y) + \ell_0^* w_0\sigma(x)\alpha(y) + \alpha^\dagger(x)\sigma(y)w_0^*\ell_0 - \alpha^\dagger(x)w_0^*\ell_0 y\\ &= \alpha^\dagger(x)w_0^*(\tilde{\sigma}(y)\ell_0 - \ell_0 y) + (\ell_0^*\tilde{\sigma}(x) - x\ell_0^*)w_0\alpha(y)\\ &= R(x,y),\end{aligned}$$

using the facts that α is a σ-derivation and that $\tilde{\sigma}(x) = w_0\sigma(x)w_0^*$. □

A triple (w_0, ℓ_0, h_0) of elements of $\mathcal{A}$ with w_0 unitary and h_0 self-adjoint will be called a <u>perturbing triple</u>. Given a restraint triple (λ, α, τ) and a perturbing triple (w_0, ℓ_0, h_0), the restraint triple $(\tilde{\lambda}, \tilde{\alpha}, \tilde{\tau})$ of Theorem 4.1 will be called the perturbation of (λ, α, τ) through (w_0, ℓ_0, h_0).

<u>Theorem</u> 4.2 Suppose that (λ, α, τ) is a restraint triple for which a corresponding quantum diffusion exists, thus there exists a family $j = (j_t : t \geq 0)$ of contractive injective homomorphisms from $\mathcal{A}$ to $\tilde{\mathcal{A}}$ such that for each $x_0 \in \mathcal{A}$, $x_t \equiv j_t(x_0)$ is an adapted process and

$$dx = \lambda(x)d\Lambda + \alpha(x)dA^\dagger + \alpha^\dagger(x)dA + \tau(x)dt. \qquad (4.6)$$

Let (w_o, ℓ_o, h_o) be a perturbing triple , and denote by w, ℓ, h respectively the adapted processes $x = (x_t : t \geq 0)$ with $x_o = w_o, \ell_o, h_o$ respectively. Suppose that there exists a unitary adapted process u satisfying the quantum stochastic differential equation

$$du = u[(w-I)d\Lambda + \ell dA^+ - \ell^+ w dA + (ih - \tfrac{1}{2}\ell^+\ell)dt], \quad u(0) = I. \tag{4.7}$$

Then a quantum diffusion exists for the perturbation $(\tilde{\lambda}, \tilde{\alpha}, \tilde{\tau})$ of (λ, α, τ) through (w_o, ℓ_o, h_o), that is there exists a family $\tilde{j} = (\tilde{j}_t : t \geq 0)$ of contractive injective homomorphisms from $\mathcal{G}$ to $\tilde{\mathcal{G}}$ such that for each $x_o \in \mathcal{G}$, $\tilde{x}_t \equiv \tilde{j}_t(x_o)$ is an adapted process and

$$d\tilde{x} = \tilde{\lambda}(\tilde{x})d\Lambda + \tilde{\alpha}(\tilde{x})dA^+ + \tilde{\alpha}^+(\tilde{x})dA + \tilde{\tau}(\tilde{x})dt. \tag{4.8}$$

In fact $\tilde{j}_t(.) = u_t j_t(.) u_t^*$; equivalently

$$\tilde{x}_t = u_t x_t u_t^*. \tag{4.9}$$

<u>Proof</u> The unitarity of u_t and the contractivity of j_t ensure that the quantum Itô formula can be used to compute the differential of the product (4.9). Using (4.6) and (4.7) and recalling the defining equations (4.1), (4.2) and (4.3) of the perturbation, it is found that $\tilde{x}$ satisfies (4.8). □

There is as yet no general existence theorem for the quantum stochastic differential equation (4.7). However in the case when λ, β and τ all vanish, that is of perturbations of the trivial (constant) diffusion, the terms w, ℓ and h in (4.7) retain their initial values, namely the ampliations of w_o, ℓ_o and h_o respectively. In this case the existence theorem of [4], Theorem 7.1 is applicable. We obtain the following corollary.

<u>Corollary</u> 4.3 Suppose that (λ, α, τ) is an admissible triple such that $\sigma = \lambda + id$ is the inner automorphism $x \longmapsto w_o x w_o^*$ where $w_o \in \mathcal{G}$ is unitary, α is the inner σ-derivation $ad_\sigma \ell_o$ and τ is given by

$$\tau(x) = i(h_o x - x h_o) - \tfrac{1}{2}(\ell_o^* \ell_o x - 2\ell_o^* \sigma(x)\ell_o + x\ell_o^*\ell_o)$$

where $h_o \in \mathcal{G}$ is self-adjoint. Then the corresponding quantum diffusion exists, being given by

$$x_t = u_t x_\circ \otimes I u_t^*$$

where u is the unique unitary process [4] satisfying

$$du = u[(w_\circ - I)\otimes Id\Lambda + \ell_\circ \otimes IdA^+ - \ell_\circ^* w_\circ \otimes IdA + (ih_\circ - \tfrac{1}{2}\ell_\circ^* \ell_\circ)\otimes Idt], \quad u_\circ = I.$$

§5. Examples

1. Diffusions on $B(\mathfrak{H}_\circ)$. Every unital endomorphism of $B(\mathfrak{H}_\circ)$ is an inner automorphism and every derivation is inner. It follows from Proposition 3.1 that evey σ-derivation is inner. Thus all quantum diffusions are described as in Corollary 4.3 by the unitary processes of [4].

2. Classical diffusions on $C^\infty(V)$. Let V be a compact Riemann manifold and let $\mathfrak{A} = C^\infty(V)$, acting on $L^2(V)$ by multiplication. Equip $\mathfrak{A}$ with the state $\omega_\circ$ of evaluation at a point $p \in V$. Let α and χ be vector fields on V, that is self-adjoint derivations of $\mathfrak{A}$. Then in view of Proposition3.4, $(0,\alpha,\tau)$ is a restraint triple, where

$$\tau(x) = \chi(x) - \tfrac{1}{2}\alpha(\alpha(x)).$$

Writing the corresponding system of stochastic differential equations as

$$dx = \alpha(x)(dA^+ + dA) + \tau(x)dt$$

and recalling [4] that $A^+ + A$ is classical Brownian motion, we obtain the algebraic description [6] of a (one-dimensional) classical diffusion on V starting at p. Multidimensional diffusion processes can be obtained similarly using several independent creation and annihilation processes.

3. Poisson processes. Let σ be an automorphism of the algebra $\mathfrak{A}$ and let $\ell_\circ$ be a non-zero element of the centre of $\mathfrak{A}$ invariant under σ. The inner σ-derivation

$$ad_\sigma \ell_\circ(x) = \ell_\circ x - \sigma(x)\ell_\circ = \ell_\circ(x - \sigma(x)) = -\ell_\circ \lambda(x)$$

is non-zero even when $\mathfrak{A}$ is commutative if $\sigma \neq id$. In view of Proposition 3.3 we may take

$$\tau(x) = -\frac{1}{2}(\ell_0^*\ell_0 x - 2\ell_0^*\sigma(x)\ell_0^* + x\ell_0^*\ell_0) = \ell_0^*\ell_0\lambda(x)$$

to obtain a restraint triple. The corresponding stochastic differential equations are

$$dx = \lambda(x)[d\Lambda - \ell dA^\dagger - \ell^* dA + \ell^*\ell\, dt] \tag{5.1}$$

but since $\lambda(\ell_0) = 0$ we may replace ℓ in (5.1) by its initial value $\ell_0 \otimes I$. The resulting equation is

$$dx = \lambda(x)\, d\Pi_{\ell_0} \tag{5.2}$$

where

$$\Pi_{\ell_0}(t) = \Lambda(t) - \ell_0 \otimes I\, A^\dagger(t) - \ell_0^* \otimes I\, A(t) + \ell_0^*\ell_0 t.$$

Π_{ℓ_0} may be interpreted as a compound Poisson process, being in particular a mutually commuting family of essentially self-adjoint operators each of whose spectra is the non-negative integers, in which the intensity is distributed according to the initial distribution of $\ell_0^*\ell_0$. The solution of (5.2) may be expressed formally as

$$x(t) = \sigma^{\Pi_{\ell_0}(t)}(x_0) \otimes I. \tag{5.3}$$

It is clear that in the commutative case the processes (5.3) will lack sample path continuity; they might therefore be thought unworthy of being called diffusions. In the non-commutative theory there is no notion of sample path, and such processes must be counted as quantum diffusions because of the unification of Brownian motion and Poisson processes brought about by quantum stochastic calculus.

<u>References</u>

[1] Accardi L, Frigerio A and Lewis J T, Quantum stochastic processes, PRIMS Kyoto <u>18</u>, 97-133(1982).

[2] Bradshaw W and Hudson R L, Quantum diffusions on the Weyl algebra, in preparation.

[3] Evans M and Hudson R L, Algebraic theory of quantum diffusions II: many dimensional diffusions, Nottingham preprint

[4] Hudson R L and Parthasarathy K R, Quantum Ito's formula and stochastic evolutions, Commun.Math.Phys.<u>93</u>, 301-323(1984)

[5] Hudson R L and Robinson P, Quantum diffusions for the non-commutative torus, in preparation.
[6] Ikeda N and Watanabe S, Stochastic differential equations and diffusion processes, North Holland (1981).
[7] Reiffel M, C^*-algebras associated with irrational rotation, Pac. J.Math.95(2)415-429(1981).

A NOTE ON INTEGRABILITY OF C^r-NORMS OF STOCHASTIC FLOWS AND APPLICATIONS

by

Yuri Kifer (*)
Institute of Mathematics
Hebrew University of Jerusalem
Jerusalem, Israel

Abstract

The integrability of uniform norms for higher derivatives of stochastic flows proved in this note implies existence of non-random stable manifolds, a Pesin type entropy formula and Hölder continuity of certain invariant subbundles for C^2-stochastic flows.

(*)Supported by Binational USA-Israel Science Foundation.

1. Introduction

Let M be a compact smooth connected m-dimensional manifold and let $Diff^r(M)$, $1 \le r \le \infty$, denote the group of C^r diffeomorphisms of M. A random process $f_t = f_t(\omega)$, $0 \le t < \infty$ defined on some probability space $(\Omega, \mathcal{F}, P)$ with values in $Diff^r(M)$ is called a stochastic flow if f_0 is the identity map id and

(i) the increments $f_{t_i} \circ f_{t_{i-1}}^{-1}, i = 1, \dots, n,$ for any $t_0 \le t_1 \le \dots \le t_n$ are independent;

(ii) for $s \le t$ the law of $f_t \circ f_s^{-1}$ depends only on $t - s$;

(iii) with probability one f_t has continuous sample paths.

If μ_t is the distribution of f_t then (i) and (ii) imply

$$\mu_t * \mu_s = \mu_{s+t} \tag{1.1}$$

i.e. $\int \phi(f) d\mu_{s+t} = \int \phi(ff) d\mu_s(f) d\mu_t(f)$.

In fact, usually it suffices to assume (1.1) in place of (i) and (ii).

Let $||f||_r$ denote the C^r norm of a diffeomorphism f (see [F]) which is defined by

$$||f||_r = \sum_{i=0}^{r} \max_j \sup_{x\in U_j} ||D^i(f\circ\Phi_j^{-1})(x)||$$

where $U_j = \Phi_j V_j$ and $\{(V_j,\Phi_j), j = 1,\ldots,el\}$ is a system of charts.

Modifying an argument from [B1] and [B2] we shall prove here the following result.

Proposition. *If* $f_t \in Diff^r(M)$, $0 \le t < \infty$ *is a stochastic flow then*

$$E((\sup_{0\le s<t} ||f_s||_r)^k + (\sup_{0\le s\le t} ||f_s^{-1}||_r)^k) < \infty \tag{1.2}$$

for any $t > 0$ *and* $k = 1,2,\ldots$ *where* E *denotes the expectation on the probability space* $(\Omega,\mathcal{F},P)$.

For $r = k = 1$ this result was proved in [B2]. The necessity of (1.2) for bigger r, probably, needs some justification. It is known in dynamical systems that results connected with regularity properties of invariant subbundles in Oseledec's multiplicative ergodic theorem, stable manifolds' theorems and Pesin's type entropy formula require more than C^1 smoothness of corresponding diffeomorphisms. In case of stochastic flows these lead to certain integrability assumptions on C^2 norms (which usually can be weakened to the integrability of $C^{1+\alpha}, \alpha > 0$ norms). These applications of (1.2) will be discussed in Section 2. In Section 3 we shall prove (1.2).

By [B1] all stochastic flows emerge essentially from stochastic differential equations. Still, the (1.2)-type integrability of uniform C^r norms can not be obtained directly from stochastic differential equations without using properties (i) - (iii) of the process f_t in $Diff^r(M)$.

2. Applications of (1.2)

The following is a version of Oseledec's multiplicative ergodic theorem for stochastic flows.

Theorem. ([C],[K]) *Let* ρ *be an invariant ergodic probability measure on* M *for a stochastic flow* f_t *i.e.* $E\rho(f_t^{-1}\Gamma) = \rho(\Gamma)$ *for any Borel* $\Gamma \subset M$. *Then for* $\rho\times P$ *almost all* (x,ω) *there exist a sequence of linear subspaces of the tangent space* T_xM *at* x,

$$0 = V^0_{(x,\omega)} \subset V^1_{(x,\omega)} \subset \dots \subset V^\nu_{(x,\omega)} = T_xM$$

and a sequence of real numbers called characteristic exponents

$$-\infty < \chi_1(\rho) < \chi_2(\rho) < \dots < \chi_\nu(\rho) < \infty$$

such that with probability one $\lim_{t\to\infty}\frac{1}{t}\ln||D_x\delta_t|| = \chi_\nu(\rho)$ *and if* $\xi \in V^i_{(x,\omega)} \setminus V^{i-1}_{(x,\omega)}$ *then* $\lim_{t\to\infty}\frac{1}{t}\ln||D_x\delta_t(\omega)\xi|| = \chi_i(\rho), i = 1,\dots,n.$ *The subspaces* $V_{(x,\omega)}$ *depend measurably on* (x,ω) *and the numbers* $d_i(\rho) = \dim V^i_{(x,\omega)} - \dim V^{i-1}_{(x,\omega)}$ *are constant* $\rho \times P$ *- almost surely (a.s).*

We formulate this theorem only for the definition of characteristic exponents and so other important assertions contained in full version of this result are skipped. A sufficient condition for this theorem is

$$\int E(\sup_{0\le t\le 1} (ln^+||D_x\delta_t|| + \ln^+||D_x\delta_t^{-1}||)d\rho(x) < \infty \tag{2.1}$$

which is much weaker than (1.2). When the relation (2.1) holds true one can prove also

Theorem. ([K]) *In the conditions of the previous theorem for* ρ*-almost all* $x \in M$ *there exists a measurable filtration of (non-random) subspaces*

$$0 = L^0_x \subset L^1_x \subset \dots \subset L^\ell_x = T_xM$$

and a sequence of (non-random) numbers

$$-\infty < \lambda_1(\rho) < \dots < \lambda_\ell(\rho) < \infty$$

such that

$$\lim_{t\to\infty}\frac{1}{t}\ln||D_x\delta_t|| = \lambda_\ell(\rho) \quad \rho \times P\text{-}a.s.$$

and if $V \in L^j_x \setminus L^{j-1}_x$ *then*

$$\lim_{t\to\infty}\frac{1}{t}\ln||D_x\delta_t V|| = \lambda_j(\rho) \quad \rho \times P\text{-}a.s.$$

The subspaces L^j_x *depend measurably on* x *and* $D_x\delta_t L^j_x = L^j_{\delta_t x}$ $\rho \times P$*-a.s.*

More delicate results require certain integrability conditions on C^2 uniform norms of the stochastic flow, for instance,

$$\int \ln^+ ||f||_2 d\mu_t(f) < \infty \quad \text{and} \quad \int \ln^+ ||f^{-1}||_2 d\mu_t(f) < \infty \tag{2.2}$$

which also follow from (1.2). Recall, that the entropy $h_\rho(f_1)$ of the random diffeomorphism f_1 with respect to an invariant measure ρ was defined in [K] as

$$h_\rho(f_1) = \sup_A \lim_{n\to\infty} \frac{1}{n} EH_\rho\left(\bigvee_{k=0}^{n-1} f_k^{-1} A\right)$$

where the supremum is taken over all finite partitions $A = \{A_1,\dots,A_\ell\}$ of M, $\bigvee_{k=0}^{n-1} f_k^{-1} A$ is the partition whose elements are $A_{i_1} \cap f_1^{-1} A_{i_2} \cap \dots \cap f_{n-1}^{-1} A_{i_n}$, $A_{i_j} \in A$, and $H_\rho(\mathcal{B}) = -\sum_{i=1}^{r} \rho(B_i) \ln \rho(B_i)$ for any finite partition $\mathcal{B} = (B_i,\dots,B_r)$,

Theorem. ([LY]). *Let* ρ *be the same as in the above theorems and* (2.2) *holds true. Suppose that* ρ *is absolutely continuous with respect to the Riemannian volume on* M. *Then*

$$h_\rho(f_1) = \sum_{i=1}^{\nu} d_i(\rho) \max(\chi_i(\rho), 0)$$

where, recall, $d_i(\rho) = \dim V^i_{(x,\omega)} - \dim V^{i-1}_{(x,\omega)}$.

The following result requires only the first part of the condition (2.2).

Theorem. [BK]. *Suppose that the transition probabilities* $P(t,x,\Gamma) = \mu_t\{f : f_x \in \Gamma\}$ *of the stochastic flow have continuous densities with respect to the Riemannian volume on* M, *and assume that* $\lambda_j(\rho) < 0$. *Then the subbundle* $\{L^j_x\}$ *is integrable in the following sense. There exists a foliation* W^j *of* $\sup \rho$ *into* C^1 *complete submanifolds* W^j_x *without boundary such that the tangent space* $T_x W^j_x$ *is* L^j_x. *For any* $x \in int \sup \rho$ *and* P*-almost every* ω *there exists a neighbourhood* $W^j_{x,\varepsilon(\omega)}$ *of* x *in* W^j_x *such that*

$$\limsup_{t\to\infty} \frac{1}{t} \ln diam(f_t(\omega) W^j_{x,\varepsilon(\omega)}) \le \lambda_j(\rho).$$

The lefthand side of (2.2) implies also the Holder continuity of subbundles L^i_x on $int \sup \rho$ and of subbundles $V_{(x,\omega)}$ is x belonging to subsets with the ρ-

measure arbitrarily close to one (see [BK]).

3. Proof of (1.2)

For any $f, g \in Diff^r(M)$ one has

$$||g \circ f||_r \le C_r ||g||_r (||f||_r^r + 1) \tag{3.1}$$

(see Lemma 3.2 in [F] where $C_r > 1$ depends only on r. Define

$$U = \{f \in Diff^r(M) : \max(||f||_r, ||f^{-1}||_r) < K\}$$

where $K = ||id||_r + 1$. Since f_t is continuous in t and $f_\circ = id$ then for any $\delta > 0$ there exists $t(\delta) > 0$ such that

$$P\{\tau_1 \le t(\delta)\} \le \delta \tag{3.2}$$

where $\tau_1 = \inf\{s : f_s \notin U\}$. Let $U^n = \{f \in Diff^r(M) : f = g_n \circ g_{n-1} \circ \dots \circ g_1$ and $g_i \in U$ for all $i = 1, \dots, n\}$
then by (3.1),

$$\max(||f||_r, ||f^{-1}||_r) \le C_r^n (K^r + 1)^n \quad \text{for } any \quad f \in U^n. \tag{3.3}$$

Put $\tau_n = \inf\{s : f_s \notin U^n\}$ then repeating the argument of Lemma 3.1 from [B1] we have

$$P\{\tau_n \le t\} = P\{f_s \notin U^n \quad \text{for } some \quad s \le t\}$$

$$\le P\{\tau_{n-1} \le t \quad \text{and} \quad f_s f_{r_{n-1}}^{-1} \notin U$$

for some s satisfying $\tau_{n-1} \le s \le t\}$

$$\le P\{\tau_{n-1} \le t\} P\{\tau_1 \le t\}$$

using the strong Markov property of the process f_t in $Diff^r(M)$ at the Markov time τ_{n-1}. Thus

$$P\{\tau_n \le t\} \le (P\{\tau_1 \le t\})^n. \tag{3.4}$$

Taking $t = t(\delta)$ and using (3.2) we get

$$P\{\tau_n \le t(\delta)\} \le \delta^n. \tag{3.5}$$

Now by (3.3) and (3.5),

$$E(\max(\sup_{0 \le s \le t(\delta)} ||f_s||_r, \sup_{0 \le s \le t(\delta)} ||f_s^{-1}||_r))^k \tag{3.6}$$

$$\le \sum_{h=1}^{\infty} C_r^{kn} (K^r + 1)^{kn} P\{\tau_{n-1} \le t(\delta) < \tau_n\}$$

$$\le \delta^{-1} \sum_{h=1}^{\infty} (C_r^k (K^r + 1)^k \delta)^n < \infty$$

if $\delta < (C_r(k^r+1))^{-k}$, where $\tau_0 = 0$. This implies (1.2) for $t \le t(\delta)$. To get (1.2) for any $t > 0$ notice that by (3.1)

$$\sup_{0\le s\le (n+1)t(\delta)} ||f_s||_r \le \sup_{0\le s\le t(\delta)} ||f_s||_r \tag{3.7}$$

$$+ \sup_{t(\delta)\le s\le (n+1)t(\delta)} ||f_s \circ f_{t(\delta)}^{-1} \circ f_{t(\delta)}||_r$$

$$\le \sup_{0\le s\le t(\delta)} ||f_s||_r$$

$$+ C_r(||f_{t(\delta)}||_r^r + 1) \sup_{t(\delta)\le s\le (n+1)t(\delta)} ||f_s \circ f_{t(\delta)}^{-1}||.$$

By (i) and (ii), $f_s \circ f_{t(\delta)}^{-1}$ is independent of $f_{t(\delta)}$ and its law is the same as for $f_{s-t(\delta)}$. Hence

$$E||f_{t(\delta)}||_r^{\ell_1} \sup_{t(\delta)\le s\le (n+1)t(\delta)} ||f_s\circ f_{t(\delta)}^{-1}||_r^{\ell_2} = E||f_{t(\delta)}||_r^{\ell_1} E \sup_{0\le s\le nt(\delta)} ||f_s||_r^{\ell_2} \tag{3.8}$$

for any $\ell_1, \ell_2 = 0,1,2,\ldots$. Thus taking $E \sup_{0\le s\le nt(\delta)} ||f_s||_r^{\ell} < \infty$ for any $\ell = 1,2,\ldots$ as an inductive assumption we obtain from (3.7) and (3.8) that

$$E \sup_{0\le s\le (n+1)t(\delta)} ||f_s||^k < \infty \quad \text{for } any \ k = 1,2,\ldots.$$

proving the integrability of the first expression in (1.2). A similar argument applied to f_s^{-1} in place of f_s completes the proof of (1.2).

REFERENCES

[B1] P. Baxendale, Brownian motions in the diffeomorphism group I, Compositio Matematica 53(1984), 19-50.

[B2] P. Baxendale, Lyapunov exponents and relative entropy for a stochastic flow of diffeomorphisms, Preprint.

[BK] M. Brin and Yu. Kifer, Dynamics of Markov chains and stable manifolds for random diffeomorphisms, Ergodic Theory of Dynamical Systems, to appear.

[C] A.P. Carverhill, Flows of stochastic dynamical systems, Ergodic theory, Stochastics 14 (1985), 273-318.

[F] J. Franks, Manifolds of C^r mappings and applications to differentiable dynamical systems, Studies in analysis, Advances in Mathematics Supplementary Series 4 (1979), 271-291.

[K] Yu. Kifer, Ergodic Theory of Random Transformations, Birkhäuser, Basel, 1986.

[LY] F. Ledrappier and L.S. Young, Entropy formula for random transformations, Preprint.

A NOTE ON STOCHASTIC MODELS WITH EXPANDING TRANSFORMATIONS

by

Yuri Kifer
Institute of Mathematics
Hebrew University of Jerusalem
Jerusalem, Israel

Abstract

Stochastic processes adapted to some expanding transformations are considered. It is proved that the trajectories of the resulting Markov chain stay "close" to non random orbits of transformations in question.

1. Introduction

Let (Ω,P) be a probability space and $\xi_0,\xi_1,\dots$ be a sequence of random vectors of the m-dimensional Euclidean space R^m.

Define $x_0 = \xi_0$ and $x_{k+1} = Ax_k + \xi_{k+1}$ for $k = 0,1,\dots,$ where A is an $m\times m$-matrix with a non-zero determinant. Set $y_k = A^{-k}x_k = \xi_0 + A^{-1}\xi_1 + \dots + A^{-k}\xi_k$. If all eigenvalues of A have absolute values greater than one and the random vectors ξ_k, $k = 0,\dots$ satisfy reasonable boundness conditions then one can easily see that the limit $y_\infty = \lim_{k\to\infty} y_k$ exists with the probability one. This means that for almost all $\omega \in \Omega$ there is a unique vector $y_\infty(\omega)$ such that the distance between $A^n y_\infty(\omega)$ and $x_n(\omega)$ cannot grow too fast as $n \to \infty$.

We shall generalize here this model so that it can be applied to certain problems of the stochastic analysis. Our approach is motivated by the related results used by the author for the study of small random perturbations of hyperbolic dynamical systems (see [4], especially §4) and by the understanding that the similar ideas work in the study of the Brownian motion on manifolds of negative curvature which gives another outlook on some problems considered in [3] and [6]. Actually, this connection was pointed out in §10 of [7].

2. Assumptions and general results

Consider a sequence of complete separable metric spaces M_n having metrics d_n together with a family of probability distributions $Q_n(x,\Gamma)$ where $x \in M_n$ and Γ is a Borel subset of M_n.

Let T_n be a sequence one to one transformations

$$T_n : M_n \to M_{n+1}$$

such that

$$d_{n+1}(T_n x, T_n y) \le q\, d_n(x,y) \tag{2.1}$$

where $q > 1$ is independent of $x, y \in M_n$ and $n = 0,1,\ldots$.

Introduce the Markov chain Θ_n with the transition probability $P_n(x,\Gamma) = Q_{n+1}(T_n x, \Gamma)$ where $x \in M_n$ and $\Gamma \subset M_{n+1}$. This means that $\Theta_n \in M_n$ and

$$P\{\Theta_{n+1} \in \Gamma \mid \Theta_n = x\} = P_n(x,\Gamma).$$

Under some assumptions on the distributions $Q_n(x,\Gamma)$ means we shall prove below that with the probability one there exists a limit

$$\Psi = \lim_{n\to\infty} T_0^{-1} T_1^{-1} T_2^{-1} \cdots T_{n-1}^{-1} \Theta_n \in M_0. \tag{2.2}$$

This means that the trajectories of the random process Θ_n stay close, in some sense, to certain trajectories of the deterministic process $x_n = T_n \cdots T_0 x$ where $x \in M_0$ depends on what random trajectory is chosen.

Theorem 1. *If for some* $r < q$,

$$\int_{M_n} d_n(x,y) Q_n(x,dy) \le C_1 r^n \tag{2.3}$$

where $C_1 > 0$ *is independent of* n *then the limit* (2.2) *exists with the probability one.*

Proof. By (2.1) one has for $k < n$ that

$$\begin{aligned} d_0(T_0^{-1}T_1^{-1} \cdots T_{k-1}^{-1}\Theta_k, T_0^{-1} \cdots T_{n-1}^{-1}\Theta_n) &\le q^{-k} d_k(\Theta_k, T_k^{-1} \cdots T_{n-1}^{-1}\Theta_n) \\ &\le q^{-k}(d_k(\Theta_k, T_k^{-1}\Theta_{k+1}) + d_k(T_k^{-1}\Theta_{k+1}, T_k^{-1}T_{k+1}^{-1}\Theta_{k+2}) \\ &\quad + \cdots + d_k(T_k^{-1} \cdots T_{n-2}^{-1}\Theta_{n-1}, T_k^{-1} \cdots T_{n-1}^{-1}\Theta_n)) \end{aligned} \tag{2.4}$$

$$\le q^{-k}\sum_{\ell=0}^{n-k-1} q^{-\ell} d_{k+\ell}(\Theta_{k+\ell}, T_{k+\ell}^{-1}\Theta_{k+\ell+1}) \le q^{-k}\sum_{\ell=1}^{n-k} q^{-\ell} d_{k+\ell}(T_{k+\ell-1}\Theta_{k+\ell-1}, \Theta_{k+\ell}).$$

By (2.3) and the Markov property it is easy to see that

$$\begin{aligned} &Ed_{k+\ell}(T_{k+\ell-1}\Theta_{k+\ell-1}, \Theta_{k+\ell}) \\ &= E\int_{M_{k+\ell}} Q_{k+\ell}(T_{k+\ell-1}\Theta_{k+\ell-1}, dy) d_{k+\ell}(T_{k+\ell-1}\Theta_{k+\ell-1}, y) \\ &\le C_1 r^{k+\ell} \end{aligned} \tag{2.5}$$

where E denotes the expectation.

$$\begin{aligned} \Pi_k &\equiv \sup_{n\ge k} d_0(T_0^{-1}\ \ldots\ T_{k-1}^{-1}\Theta_k, T_0^{-1}\ \ldots\ T_{n-1}^{-1}\Theta_n) \\ &\le q^{-k}\sum_{\ell=1}^{\infty} q^{-\ell} d_{k+\ell}(T_{k+\ell-1}\Theta_{k+\ell-1}, \Theta_{k+\ell}) \end{aligned} \tag{2.6}$$

and so by (2.5),

$$E\Pi_k \le C_1 (\tfrac{r}{q})^{-k}(1 - \tfrac{r}{q})^{-1}. \tag{2.7}$$

Hence by the Chebyshev's inequality for any $\varepsilon > 0$,

$$P\{\Pi_k \ge \varepsilon\} \le C_2 \varepsilon^{-1}(\tfrac{r}{q})^{-k} \tag{2.8}$$

where $C_2 = C_1(1 - \frac{r}{q})^{-1}$.

Since $r < q$ then

$$\sum_{k=0}^{\infty} P\{\Pi_k \ge \varepsilon\} < \infty \tag{2.9}$$

and so by the Borel-Cantelli lemma for any ε and almost every $\omega \in \Omega$ there exists $k(\varepsilon,\omega)$ such that if $k > k(\varepsilon,\omega)$ then $\Pi_k(\omega) < \varepsilon$. By the Cauchy criteria this means that the limit (2.2) exists almost sure.

Remark 1. If $M_n \equiv M_0$ and $T_n \equiv T_0$ for all $n = 1,\ldots$ then one can prove (2.2) under some kind of Kesten's condition (see [2]) in the form

$$P\{d(T\Theta_n, \Theta_{n+1}) \ge \alpha d(O,\Theta_n)^{1-2\delta} / \Theta_0,\ldots,\Theta_n\} \le K\alpha^{-2}$$

for all $O \le d \le \tilde{K} d(O,\Theta_n)^{2\delta}$ where O is the fixed point of the transformation T.

The following result gives more precise estimates of the asymptotic behaviour of Θ_n.

Theorem 2. *Let* $T_0, T_1, T_2, \ldots$ *be continuous and*

$$\int_{M_n} (d_n(x,y))^{\gamma} Q_n(x,dy) \le C_3 \tag{2.10}$$

for some $\gamma, C_3 > 0$ *independent of* $n = 0,1,\ldots$. *Then for any* $\kappa > 0$ *there exists a random variable* D_κ *such that with the probability one*

$$d_k(\Theta_k, T_{k-1} \cdots T_0 \Psi) \le D_\kappa k^{(1+\kappa)\gamma^{-1}} \tag{2.11}$$

for all $k = 0,1,\ldots,$ *where* Ψ *is given by* (2.2) *according to Theorem* 1.

Proof. Set

$$\Psi_n = T_0^{-1} \cdots T_{n-1}^{-1} \Theta_n \tag{2.12}$$

then by (2.1)

$$\begin{aligned} d_k(\Theta_k, T_{k-1} \cdots T_0 \Psi_n) &= d_k(\Theta_k, T_k^{-1} \cdots T_{n-1}^{-1} \Theta_n) \\ &\le \sum_{\ell=0}^{n-k-1} d_k(T_k^{-1} \cdots T_{k+\ell-1}^{-1} \Theta_{k+\ell}, T_k^{-1} \cdots T_{k+\ell}^{-1} \Theta_{k+\ell+1}) \\ &\le \sum_{\ell=1}^{n-k} q^{-\ell} d_{k+\ell}(T_{k+\ell-1} \Theta_{k+\ell-1}, \Theta_{k+\ell}). \end{aligned} \tag{2.13}$$

From (2.10) and the Chebyshev's inequality it follows that

$$P\{d_m(T_{m-1}\Theta_{m-1}, \Theta_m) \ge m^{(1+\kappa)\gamma^{-1}}\} \le C_3 m^{-(1+\kappa)}. \tag{2.14}$$

Since $\sum_{m=1}^{\infty} m^{-(1+\kappa)} < \infty$ then by the Borel-Cantelli lemma there exists a random variable M_κ such that

$$d_m(T_{m-1}\Theta_{m-1}, \Theta_m) < m^{(1+\kappa)\gamma^{-1}} \quad \text{if} \quad m \ge M_\kappa \tag{2.15}$$

with the probability one.

Put

$$R_\kappa = \max(1, \max_{1 \le m \le M_\kappa} d_m(T_{m-1}\Theta_{m-1}, \Theta_m)) \tag{2.16}$$

then (2.15) gives

$$d_m(T_{m-1}\Theta_{m-1}, \Theta_m) \le R_\kappa m^{(1+\kappa)\gamma^{-1}} \quad \text{for } all \quad m = 1,2,\ldots \tag{2.17}$$

with the probability one. This together with (2.13) yield

$$d_k(\Theta_k,T_{k-1}\,\cdots\,T_0\Psi_n) \geq R_\kappa \sum_{\ell=1}^{n-k} q^{-\ell}(k+\ell)^{(1+\kappa)\gamma^{-1}} \leq D_\kappa k^{(1+\kappa)\gamma^{-1}} \tag{2.18}$$

where

$$D_\kappa = R_\kappa \sum_{\ell=1}^{\infty} q^{-\ell}(1+\ell)^{(1+\kappa)\gamma^{-1}}. \tag{2.19}$$

Since $T_0,\ldots,T_{k-1}$ are continuous transformations and by Theorem 1 $\Psi_n \to \Psi$ as $n \to \infty$ in the space M_0 then

$$d_k(\Theta_k,T_{k-1}\,\cdots\,T_0\Psi_n) \to d_k(\Theta_k,T_{k-1}\,\cdots\,T_0\Psi) \tag{2.20}$$

as $n \to \infty$. The right hand side of (2.18) is independent of n that together with (2.20) imply (2.11) proving Theorem 2.

Theorem 3. *If* $T_0,T_1,\ldots$ *are continuous transformations and for some* $r>0$

$$Q_n(x,\{y : d_n(x,y) \leq r\}) = 1 \tag{2.21}$$

for all $x \in M_n$ *and* $n = 0,1,\ldots$ *then with the probability one*

$$d_k(\Theta_k,T_{k-1}\,\cdots\,T_0\Psi) \leq (1-q)^{-1}r \tag{2.22}$$

for all $k = 1,\ldots$ *where* Ψ *is given by* (2.2).

Proof. We use (2.13) taking into account that

$$d_{k+\ell}(T_{k+\ell-1}\Theta_{k+\ell-1},\Theta_{k+\ell}) \leq r$$

with the probability one. Then

$$d_k(\Theta_k,T_{k-1}\,\cdots\,T_0\Psi_n) \leq r\sum_{\ell=1}^{\infty} q^{-\ell} = r(1-q)^{-1}$$

proving (2.22).

3. Applications

The first example is connected with the Brownian motion on a complete, connected, simply connected manifold X having sectional curvatures bounded between two negative constants. We shall assume that absolute values of the derivatives of sectional curvatures are also bounded by some constant.

Let M_n be a geodesic sphere of the radius n with the centre at some fixed point x_0. Then in the geodesic spherical coordinates the Laplacian which is the generator of the process B_t has the form

$$\Delta = \frac{\partial^2}{\partial\rho^2} + Y(\rho)\frac{\partial}{\partial\rho} + L_{s,\rho} \tag{3.1}$$

where $Y(\rho) > a$ for some constant $a > 0$ independent of ρ and L_s is the operator with bounded coefficients including the derivatives in tangential directions to the sphere M.

Let $\tau_n = \inf\{t : B_t \in M_n\}, n = 1,2,\ldots.$ Then it is easy to see using the comparison theorems from the theory of partial differential equations in the same way as in Lemma 2 of [3] that

$$P_y\{\tau_{n+1} > t\} \le C_4 e^{-\beta t} \tag{3.2}$$

for some $C_4, \beta > 0$ independent of $y \in M_n$. Since the operators $L_{s,\rho}$ have uniformly bounded coefficients then it follows from (3.2) that

$$E_y(d_{n+1}(T_n y, B_{\tau_{n+1}}))^k \le C^{(k)} < \infty \tag{3.3}$$

where $C^{(k)} > 0$ depends only on $k = 1,2,\ldots$ but not on $y \in M_n$ or $n = 1,2,\ldots$ and T_n maps M_n on M_{n+1} in the following way: if $y \in M_n$ then $T_n y$ is the point of the intersection of M_{n+1} with the geodesic passing through x_0 and y.

It is easy to see now that all assumptions of Theorem 2 are satisfied for any $\gamma > 0$. Hence we conclude that for almost all paths of B_{τ_n} one can find a geodesic such that the distance between B_{τ_n} and this geodesic (considered as a set) grows slower than any power of n. The same result can be proved for B_t itself.

Now consider the next application of our theory. Let x_t be a diffusion process in R^m having the generator of the form

$$\mathrm{L} = \mathrm{L}_0 + (B,\nabla) \tag{3.4}$$

where L_0 is a second order elliptic operator with bounded smooth coefficients and (B,∇) is the first order operator generating a dynamical system S^t by the formula

$$\frac{d(S^t x)}{dt} = B(S^t x). \tag{3.5}$$

Assume that for some $t_0 > 0$ and $q > 1$,

$$|S^{t_0}x - S^{t_0}y| \ge q|x - y| \tag{3.6}$$

for any $x, y \in R^m$.

Now one can easily check the conditions of Theorem 2 to obtain that with the probability one there exists a limit

$$\Psi = \lim_{t\to\infty} S^{-t} x_t \tag{3.7}$$

and moreover that x_t stays closer to $S^t\Psi$ than any power of t.

This result allows to investigate some boundaries for the process x_t in the spirit of [1].

Sometimes our approach can be used for calculating the escape rates as defined in [5]. For this we need to know the asymptotic of $t^{-1}\log P_x\{x_t \in K\}$ where $x \in R^m$, K is a compact set and we consider the diffusion process x_t generated by the operator L satisfying (3.4) - (3.6).

If one can show that $\Psi_t = S^{-t}x_t$ have distributions with densities bounded by a constant $C_5 > 0$ independent of t then the asymptotics in question is the same as for $t^{-1}\log(\textit{volume}\ S^{-t}K)$. This last asymptotic behaviour can be easily studied. Indeed, since S^t is expanding transformation of R^m then it has a unique fixed point O. Let $B(x) = \Pi x + O(|x|^2)$ where Π is a matrix. Clearly, all eigenvalues of Π have positive real parts which sum we denote by Λ. Then it is easy to see that

$$\lim_{t\to\infty} t^{-1}\log(\textit{volume}\ S^{-t}K) = -\Lambda.$$

4. Hyperbolic systems

Let X_n and Y_n, $n = 0,1,\ldots$ be two sequences of metric spaces with the metrics ρ_n and r_n, respectively. Consider one to one transformations

$$R_n : X_n \to X_{n+1} \quad \textit{and} \quad S_n : Y_n \to Y_{n+1}$$

satisfying

$$\begin{aligned} &\rho_{n+1}(R_n x_1, R_n x_2) \ge q\rho_n(x_1,x_2) \quad \textit{and} \\ &r_{n+1}(S_n y_1, S_n y_2) \le q^{-1} r_n(y_1,y_2) \end{aligned} \tag{4.1}$$

for some $q > 1$ independent of $x_1,x_2 \in X_n$ and $y_1,y_2 \in Y_n$.

Introduce the product space $Z_n = X_n \times Y_n$ with the metric

$$d_n(x_1\times y_1, x_2\times y_2) = \rho_n(x_1,x_2) + r_n(y_1,y_2) \tag{4.2}$$

for any $x_1,x_2 \in X_n$ and $y_1,y_2 \in Y_n$. Define the transformations $T_n : Z_n \to Z_{n+1}$ acting by the formula $T_n x\times y = R_n x \times S_n y$.

Let $Q_n(z,\Gamma)$ be a family of probability distributions on Z_n with $z \in Z_n$ and Γ being a Borel subset of Z_n. Consider the Markov chain Θ_n with the transition probability $P_n(z,\Gamma) = Q_{n+1}(T_n z,\Gamma)$ where $z \in Z_n$ and $\Gamma \subset Z_{n+1}$.

In these circumstances one can prove the results similar to Theorems 1-3 using the same method as in §2.

Theorem 4. *Let* $\Theta_n = \Theta_n^x \times \Theta_n^y$ *be the unique decomposition with* $\Theta_n^x \in X_n$ *and* $\Theta_n^y \in Y_n$.

a) If $\int_{Z_n} d_n(v,w) Q_n(v,dw) \le C_6 r^n$ *for some* $r < q$ *then the limits*

$\psi^x = \lim_{n\to\infty} T_0^{-1} \dots T_{n-1}^{-1} \Theta_n^x$ *and* $\psi^y = \lim_{n\to\infty} \Theta_n^y$ *exist with the probability one;*

b) If $\int_{Z_n} (d_n(v,w))^\gamma Q_n(v,dw) \le C_7$, *and* $T_0, T_1, \dots$ *are continuous then for any* $x > 0$ *there exists a random variable* D_κ *such that*

$$d_k(\Theta_k^x, T_{k-1} \dots T_0 \psi^x) \le D_\kappa k^{(1+\kappa)\gamma^{-1}}$$

for all $k = 0,1,\dots$.

c) If $T_0, T_1, \dots$ *are continuous and for some* $r > 0$,

$$Q_n\{v, \{w : d_n(v,w) \le r\}) = 1$$

for all $v \in Z_n$ *and* $n = 0,1,\dots$ *then with the probability one*

$$d_k(\Theta_k, (T_{k-1} \dots T_0 \psi^x) \times \psi^y) \le 2(1-q)^{-1} r$$

for all $k = 1,\dots$.

REFERENCES

[1] M. Cranston, S. Orey, U. Rosler, Exterior Dirichlet problem and the asymptotic behavior of diffusions, in: Lecture Notes in Control and Information Sciences 25(1978), 207-220, Springer, Berlin.

[2] H. Kesten, Some nonlinear stochastic growth models, Bull. Amer. Math. Soc. 77(1971), 492-511.

[3] Yu. Kifer, Brownian motion and harmonic functions on manifolds of negative curvature, Theor. Prob. Appl. 2(1976), 81-95.

[4] Yu. Kifer, General random perturbations of hyperbolic and expanding transformations, J. D'Analyse Math. 47(1986).

[5] M. Pinsky, Large deviations for diffusion process in: Stochastic Analysis, (1978), 271-283.

[6] M.J.-J. Prat, Etude asymptotique du mouvement brownien sur une variété riemannienne à courbure negative, C.R. Acad. Sci. Ser. A, 272(1971), 1586-1589.

[7] Ja.G. Sinai, Gibbs measures in ergodic theory, Russian Math. Surveys 27(1972), No. 4, 21-70.

The Large Deviation Principle in Statistical Mechanics: an Expository Account

J.T. Lewis
Dublin Institute for Advanced Studies
10 Burlington Road
Dublin 4, Ireland

§1 Introduction

At the 1983 Swansea meeting I described joint work with J.V. Pulè [1] on the weak law of large numbers in statistical mechanics; at that time we were unaware of its connection with the principle of large deviations as formulated by Varadhan [2]. Since then, we have found Varadhan's theorems to be a powerful tool in statistical mechanics (results of use in investigating models of an interacting boson gas are described elsewhere in this volume [3]) and so it may prove worthwhile to look again at the results of [1] with the benefit of hindsight.

§2 The Grand Canonical Pressure

Since the pioneering work of van Hove [4], the importance of proving the existence of thermodynamic functions in the thermodynamic limit has been recognized. We recall the definition of the grand canonical pressure: consider a sequence $\{\Lambda_\ell : \ell = 1, 2, \ldots\}$ of regions of Euclidean space R^d, and denote the volume of Λ_ℓ by V_ℓ; associated with each region Λ_ℓ is a countable set Ω_ℓ, the set of configurations of particles in Λ_ℓ; on Ω_ℓ are defined random variables $\mathrm{H}_\ell : \Omega_\ell \to \mathrm{R}$ and $\mathrm{N}_\ell : \Omega_\ell \to \mathrm{N}$; $\mathrm{H}_\ell(\omega)$ is interpreted as the energy of the configuration ω and $\mathrm{N}_\ell(\omega)$ as the number of particles in ω. The grand canonical measure P_ℓ^μ with chemical potential μ is defined on subsets of Ω_ℓ by

$$\mathrm{P}_\ell^\mu [A] = \Xi_\ell(\mu)^{-1} \sum_{\omega \epsilon \mathrm{A}} e^{\beta\{\mu \mathrm{N}_\ell(\omega) - \mathrm{H}_\ell(\omega)\}}; \qquad (2.1)$$

here $\beta = 1/k\mathrm{T}$ is the inverse temperature and $\Xi_\ell(\mu)$ is the grand canonical partition function given by

$$\Xi_\ell(\mu) = \sum_{\omega \epsilon \Omega_\ell} e^{\beta\{\mu \mathrm{N}_\ell(\omega) - \mathrm{H}_\ell(\omega)\}}. \qquad (2.2)$$

The grand canonical pressure $\mathrm{p}_\ell(\mu)$ is defined by

$$\mathrm{p}_\ell(\mu) = (\beta \mathrm{V}_\ell)^{-1} \ln \Xi_\ell(\mu). \qquad (2.3)$$

It is closely related to the cumulant generating function for the particle number density $X_\ell = N_\ell/V_\ell$; a straightforward calculation yields the formula

$$\int_{[0,\infty)} e^{\beta V_\ell xt} K_\ell^\mu[dx] = e^{\beta V_\ell\{p_\ell(\mu+t)-p_\ell(\mu)\}}, \tag{2.4}$$

where K_ℓ^μ is the distribution function for X_ℓ defined by $K_\ell^\mu = P_\ell^\mu \circ X_\ell^{-1}$. We have introduced the concepts associated with the grand canonical pressure in the simplest case, namely, when Ω_ℓ is a countable set. But the formula (2.4), linking the distribution function K_ℓ^μ with the grand canonical pressure holds in wider contexts; for example, when Ω_ℓ is an arbitrary measure space carrying a pair of random variables H_ℓ and N_ℓ from which a grand canonical Gibbs measure can be defined, or when H_ℓ and N_ℓ are commuting self-adjoint operators on some hilbert space $\mathcal{H}_\ell$ such that trace $(e^{\beta\{\mu N_\ell - H_\ell\}})$ is finite. Our first assumption in the remainder of this note is that the distribution function K_ℓ^μ satisfies (2.4) for some function $p_\ell(\mu)$. Our second assumption is that $p(\mu) = \lim_{\ell\uparrow\infty} p_\ell(\mu)$ exists. From these two assumptions, much follows: the large deviation upper bound holds and, with it, the Berezin-Sinai criterion for a first-order phase-transition; the large deviation lower bound holds in the complement of first-order phase-transition segments; if the limit function $\mu \mapsto p(\mu)$ is differentiable at some μ_o then $\{K_\ell^{\mu_o} : \ell = 1, 2, ...\}$ converges in distribution to the degenerate distribution concentrated at $p'(\mu_o)$.

In probability theory, it is natural to prove large deviation results for sums of independent, or weakly dependent, random variables; the Markov chain condition is an example of a condition of weak dependence. I claim that, in statistical mechanics, the natural condition of weak dependence is the existence of the pressure. In this note, we explore the consequences for the distribution of the particle number density of the existence of the pressure.

§3 The General Setting

For each μ in some open interval D of the real line, let $\{K_\ell^\mu : \ell = 1, 2, ...\}$ be a sequence of probability measures on $[0, \infty)$ satisfying

(P1)

$$\int_{[0,\infty)} e^{V_\ell tx} K_\ell^\mu[dx] = e^{V_\ell\{p_\ell(\mu+t)-p_\ell(\mu)\}} < \infty,$$

where $\{V_\ell : \ell = 1, 2, ...\}$ is a sequence of positive constants diverging to $+\infty$.

(P2) The limit $p(\mu) = \lim_{\ell\uparrow\infty} p_\ell(\mu)$ exists for all values of μ in the interval of definition, D.

The first consequence makes use of Hölder's Inequality and we omit the proof:

Lemma 1

Assume that (P1) holds; then $\mu \mapsto p_\ell(\mu)$ is convex. Assume, in addition, that (P2) holds; then $\mu \mapsto p(\mu)$ is convex.

Lemma 2

For all μ and $\mu + \alpha$ in the domain of definition of K_ℓ, the measures K_ℓ^μ and $K_\ell^{\mu+\alpha}$ are mutually absolutely continuous:

$$K_\ell^{\mu+\alpha}[dx] = e^{V_\ell c_\ell^\mu(x;\alpha)} K_\ell^\mu[dx], \tag{3.1}$$

where

$$c_\ell^\mu(x;\alpha) = \alpha x + p_\ell(\mu) - p_\ell(\mu+\alpha). \tag{3.2}$$

Proof:

$$\int_{[0,\infty)} e^{xt} \,.\, e^{V_\ell \alpha x} K_\ell^\mu[dx] = e^{V_\ell\{p_\ell(\mu+\alpha+\frac{t}{V_\ell}) - p_\ell(\mu)\}},$$

by (P1); again, by (P1), we have

$$\frac{e^{V_\ell p_\ell(\mu+\alpha)}}{e^{V_\ell p_\ell(\mu)}} \int_{[0,\infty)} e^{xt} K_\ell^{\mu+\alpha}[dx]$$

$$= e^{V_\ell\{p_\ell(\mu+\alpha+\frac{t}{V_\ell}) - p_\ell(\mu)\}}.$$

The claim follows from the uniqueness theorem for Laplace transforms.

Theorem 1

Assume that (P1) and (P2) hold and that p is differentiable at μ; then
(1) the limit $\rho = \lim_{\ell\uparrow\infty} \int_{[0,\infty)} x\, K_\ell^\mu[dx]$ exists.
(2) the sequence $\{K_\ell^\mu : \ell = 1, 2, \ldots\}$ converges weakly to the degenerate distribution δ_ρ concentrated at ρ.

The proof utilises the convexity of the functions p_ℓ, $\ell = 1, 2, \ldots$, established in Lemma 1. This enables us to apply

Griffith's Lemma

Let $\{f_\ell : \ell = 1, 2, ...\}$ be a sequence of convex functions defined on a common open interval G converging pointwise to a function f. Let $\{x_\ell : \ell = 1, 2, ...\}$ be a sequence of points of G converging to a point x of G. Then

$$f'_-(x) \leq \liminf_{\ell\to\infty} (f_\ell)'_-(x_\ell) \leq \limsup_{\ell\to\infty} (f_\ell)'_+(x_\ell) \leq f'_+(x).$$

(See [1] and references contained therein.)

Proof of Theorem 1:

By an elementary computation, $\int_{[0,\infty)} x\,K^\mu_\ell[dx] = p'_\ell(\mu)$ since, by (P1), the moment generating function $s \mapsto \int_{[0,\infty)} e^{sx}\,K^\mu_\ell[dx]$ of K^μ_ℓ is finite on a neighbourhood of zero. Since p is assumed to be differentiable at μ, it follows from Griffith's Lemma that $\{p'_\ell(\mu) : \ell = 1, 2, ...\}$ converges to $p'(\mu)$since, by(P2), $\{p_\ell : \ell = 1, 2, ..\}$ converges pointwise to p on D. Thus (1) holds with $\rho = p'(\mu)$.

By (P1), we have

$$\int_{[0,\infty)} e^{sx}\,K^\mu_\ell[dx] = e^{s\{p_\ell(\mu+\frac{s}{V_\ell})-p_\ell(\mu)\}/(\frac{s}{V_\ell})} \tag{3.3}$$

for s in a neighbourhood of zero and hence for all s in $[-\infty, 0]$. Fix s and put $\mu_\ell = \mu + \frac{s}{V_\ell}$; for ℓ sufficiently large, μ_ℓ is in D; moreover, $\lim_{\ell\uparrow\infty} \mu_\ell = \mu$. By the convexity of $\mu \mapsto p_\ell(\mu)$, we have

$$(p_\ell)'_+(\mu) \leq \{p_\ell(\mu + \frac{s}{V_\ell}) - p_\ell(\mu)\}/(\frac{s}{V_\ell}) \leq (p_\ell)'_-(\mu_\ell). \tag{3.4}$$

Since p is differentiable at μ, it follows from Griffith's Lemma that both $\{(p_\ell)'_+(\mu)\}$ and $\{(p_\ell)'_-(\mu_\ell)\}$ converge to $p'(\mu)$. Thus we have

$$\lim_{\ell\to\infty} \int_{[0,\infty)} e^{sx}\,K^\mu_\ell[dx] = e^{s\rho}; \tag{3.5}$$

But $\int_{[0,\infty)} e^{sx}\delta_\rho[dx] = e^{s\rho}$, so that (2) follows by the continuity and uniqueness theorems for the Laplace transform.

Thus we have established that if the pressure p exists and is differentiable at μ then the sequence $\{K^\mu_\ell : \ell = 1, 2, ..\}$ satisfies the weak law of large numbers.

§4 The Heuristics of Large Deviations

Returning to the context of §2, we have the following reformulation of Theorem 1:

Suppose that the pressure $p = \lim_{\ell\uparrow\infty} p_\ell$ exists pointwise on the interval D on which the p_ℓ are defined and that p is differentiable at μ; then

(1) the limit $\rho = \lim_{\ell\uparrow\infty} E_\ell^\mu[X_\ell]$ exists.
(2) Let $g : [0,\infty) \to R$ be a bounded function which is continuous at ρ; then $\lim_{\ell\uparrow\infty} E_\ell^\mu[g(X_\ell)] = g(\rho)$.

Proof: (1) is a straight translation:

$$E_\ell^\mu[X_\ell] = \sum_{\omega\,\epsilon\,\Omega} X_\ell(\omega)P_\ell^\mu[\omega] = \int_{[0,\infty)} x\,K_\ell^\mu[dx].$$

To prove (2), choose $\epsilon > 0$; by the continuity of g at ρ, there exists a neighbourhood I_ρ of ρ on which $|\,g(x) - g(\rho)| < \epsilon$. Now

$$g(\rho) - E_\ell^\mu[g(X_\ell)] = \int_{[0,\infty)} (g(\rho) - g(x))K_\ell^\mu[dx]. \tag{4.1}$$

Thus

$$|g(\rho) - E_\ell^\mu[g(X_\ell)]| \le \int_{I_\rho} |g(\rho) - g(x)|K_\ell^\mu[dx]$$

$$+\int_{I_\rho^c} |g(\rho) - g(x)|K_\ell^\mu[dx] \le \epsilon + 2M\,K_\ell^\mu[I_\rho^c], \tag{4.2}$$

where $M = \sup_{x\epsilon[0,\infty)} g(x)$. But, by Theorem 1, $K_\ell^\mu \to \delta_\rho$; this means that there exists ℓ_o such that, for all $\ell > \ell_o$, we have $K_\ell^\mu[I_\rho^c] < \epsilon$. Hence, for all $\ell > \ell_o$,

$$|g(\rho) - E_\ell^\mu[g(X_\ell)]| \le \epsilon\,(1 + 2M); \tag{4.3}$$

but ϵ was an arbitrary positive number, so that

$$\lim_{\ell\to\infty} E_\ell^\mu[g(X_\ell)] = g(\rho) \tag{4.4}$$

It sometimes happens in statistical mechanics that we find it interesting to introduce a perturbed grand canonical measure $\tilde{P}_\ell^\mu$ which is conveniently defined via its expectation functional $\tilde{E}_\ell^\mu[\,\cdot\,]$:

$$\tilde{E}_\ell^\mu[A] = \frac{E_\ell^\mu[A e^{\beta V_\ell u(X_\ell)}]}{E_\ell^\mu[e^{\beta V_\ell u(X_\ell)}]}, \tag{4.5}$$

where u is a continuous function on $[0,\infty)$ which is bounded above. Can we say anything about

$$\lim_{\ell\to\infty} \tilde{E}_\ell^\mu[g(X_\ell)]?$$

If the V_ℓ factor were absent from the exponent, we could conclude that the limit would be the same as before: $g(\rho)$. However, the presence of the factor V_ℓ causes the fluctuations in X_ℓ to contribute to the limiting value; we expect that the answer will be $g(\tilde{\rho})$ where $\tilde{\rho} \neq \rho$, in general. There are two ways of proving this; they are closely related, as we might expect.

We can introduce a perturbed Hamiltonian $\tilde{H}_\ell(\omega) = H_\ell(\omega) + V_\ell(u \circ X_\ell)(\omega)$ and use it to define a perturbed pressure $\tilde{p}_\ell(\mu)$. A straightforward manipulation gives

$$\tilde{p}_\ell(\mu) = p_\ell(\mu) + \frac{1}{\beta V_\ell} \ln E_\ell^\mu[e^{\beta V_\ell u(X_\ell)}]. \tag{4.6}$$

Since

$$E_\ell^\mu[e^{\beta V_\ell u(X_\ell)}] = \int_{[0,\infty)} e^{\beta V_\ell u(x)} K_\ell^\mu[dx], \tag{4.7}$$

proof of the existence of the pressure

$$\tilde{p}(\mu) = \lim_{\ell\to\infty} \tilde{p}_\ell(\mu)$$

amounts to proving the existence of the limit

$$\lim_{\ell\to\infty} \frac{1}{\beta V_\ell} \ln \int_{[0,\infty)} e^{\beta V_\ell u(x)} K_\ell^\mu[dx];$$

conditions on $\{K_\ell^\mu : \ell = 1, 2, ...\}$ sufficient to ensure this were given by Varadhan [2] in a general setting, and we will give a precise statement of them in the next section. Roughly speaking, they are that there exists a function $I^\mu(\cdot) : [0.\infty) \to [0,\infty]$ such that $K_\ell^\mu[dx] \sim e^{-\beta V_\ell I^\mu(x)} dx$; in our case, the only zero of I^μ is at $x = \rho = p'(\mu)$, so that $I^\mu(\cdot)$ determines the rate at which $P_\ell^\mu[A]$ goes to zero if ρ is not in A. Intuitively, one would expect that

$$\lim_{\ell\to\infty} \frac{1}{\beta V_\ell} \ln \int_{[0,\infty)} e^{\beta V_\ell u(x)} K_\ell^\mu[dx] = \sup_{[0,\infty)} \{u(x) - I^\mu(x)\}; \tag{4.8}$$

this is the conclusion of Varadhan's First Theorem. It follows that the perturbed pressure $\tilde{p}(\mu)$ exists and is given by

$$\tilde{p}(\mu) = p(\mu) + \sup_{[0,\infty)} \{u(x) - I^\mu(x)\}, \tag{4.9}$$

provided that u and $\{K_\ell^\mu\}$ satisfy the hypotheses of the theorem. If $\tilde{p}$ is differentiable at μ, it follows from our previous argument that

$$\lim_{\ell\to\infty} \tilde{E}_\ell^\mu[g(X_\ell)] = g(\tilde{\rho}) \tag{4.10}$$

where now $\tilde{\rho} = \tilde{p}'(\mu)$.

Another way of computing this limit is via Varadhan's Second Theorem: if the supremum $\sup_{[0,\infty)} \{u(x) - I^\mu(x)\}$ is attained at an isolated point x^*, then

$$\lim_{\ell\to\infty} \tilde{E}_\ell^\mu[g(X_\ell)] = g(x^*).$$

We shall see, in our case, that the supremum is attained at an isolated point if and only if $\tilde{p}$ is differentiable at μ and then $x^* = \tilde{p}'(\mu)$.

We have seen that large deviations from the mean (deviations on the scale of V_ℓ) are of importance in the evaluation of

$$\lim_{\ell\to\infty} \frac{E_\ell^\mu[g(X_\ell)e^{\beta V_\ell u(X_\ell)}]}{E_\ell^\mu[e^{\beta V_\ell u(X_\ell)}]}.$$

It is for this reason that we are interested in the rate at which the degenerate distribution is approached; those distributions which approach the degenerate distribution exponentially fast are said to satisfy the large deviation principle. Next, we turn to the precise definition of this concept.

§5 Varadhan's Theorems

Donsker initiated the study of singular perturbations of partial differential equations by means of functional integration; he showed how, in the case of Burger's equation, a transformation introduced by Hopf can be used to convert the equation to a linear equation which can be solved as a function space integral. The perturbation problem can then be studied by an analysis of the asymptotic behaviour of function space integrals; in the case of Burger's equation, the asymptotic analysis was carried out by Schilder [5]. Varadhan [2] showed how a class of such problems can be treated using more general families of measures on function space whose asymptotic behaviour is to be investigated; in §3 of [2], Varadhan gave an account of the asymptotic analysis in an abstract setting. Subsequently, in a sequence of papers, Donsker and Varadhan applied these methods to a wide variety of problems involving stochastic processes (a full bibliography can be found in Varadhan's monograph [6]). The method is a far-reaching generalization of the saddle-point method or Laplace's method for one-dimensional integrals. It is our experience that whenever, in statistical mechanics, an author claims to use the saddle-point method, an efficient way of giving a rigorous proof is to check that the hypotheses of Varadhan's Theorems are verified. For that reason, we summarize here the results proved in §3 of [2].

Let E be a complete separable metric space; let $\{K_\ell : \ell = 1, 2, ...\}$ be a sequence of probability measures on the σ - field of Borel subsets of E and let $\{V_\ell : \ell = 1, 2, ...\}$ be a sequence of non-negative numbers such that $V_\ell \to \infty$. We say that $\{K_\ell\}$ obeys the large deviation principle with constants $\{V_\ell\}$ and rate-function $I(\cdot)$ if there exists a function $I : E \to [0, \infty]$ satisfying:

(LD1): $I(\cdot)$ is lower semi-continuous on E.

(LD2): For each finite m, $\{x : I(x) \leq m\}$ is compact.

(LD3): For each closed subset C of E,
$$\limsup_{\ell\to\infty} \frac{1}{V_\ell} \ln K_\ell[C] \leq - \inf_{C} I(x).$$

(LD4): For each open subset G of E,
$$\liminf_{\ell\to\infty} \frac{1}{V_\ell} \ln K_\ell[G] \geq - \inf_{G} I(x).$$

For example, if $I(\cdot)$ is a lower semi-continuous function whose level sets are compact and m is a σ - finite measure on E such that $x \to e^{-I(x)}$ is integrable with respect to m, and $\{V_\ell\}$ is a sequence of non-negative numbers such that $V_\ell \to \infty$, then the sequence $\{K_\ell\}$ of probability measures defined by

$$K_\ell[A] = \frac{\int_A e^{-V_\ell I(x)} m(dx)}{\int_E e^{-\beta V_\ell I(x)} m(dx)} \tag{5.1}$$

satisfies the large deviation principle with constants $\{V_\ell\}$ and rate-function $I(\cdot)$. The definition above has the advantage that it does not require the existence of a reference measure such as m. We are now in a position to state

Varadhan's First Theorem

Let $\{K_\ell : \ell = 1, 2, ...\}$ be a sequence of probability measures on E obeying the large deviation principle with constants $\{V_\ell\}$ and rate-function $I(\cdot)$. Then, for any continuous function G on E which is bounded above, we have

$$\lim_{\ell\to\infty} \frac{1}{V_\ell} \ln \int_{[0,\infty)} e^{V_\ell G(x)} K_\ell[dx] = \sup_{E} \{G(x) - I(x)\}.$$

The condition that G be bounded above can be weakened; it is enough to suppose that $\sup\{G(x) : x \in \cup_{\ell\geq 1} \operatorname{supp} K_\ell\}$ is finite. The theorem can be extended to cover the situation where the function G is replaced by a sequence of functions $\{G_\ell : \ell = 1, 2, \ldots\}$; this is Theorem 3.4 of [2].

Let $\tilde{K}_\ell$ be defined by

$$\tilde{K}_\ell[A] = \int_A e^{V_\ell G(x)} K_\ell[dx] / \int_E e^{V_\ell G(x)} K_\ell[dx];$$

Varadhan's Second Theorem gives sufficient conditions for the sequence of perturbed measures to converge weakly to a degenerate distribution.

Varadhan's Second Theorem

Suppose that $\Lambda = \sup_E \{G(x) - I(x)\}$ is attained at a point x^* of E and that

$$\sup_{\{x : d(x,x^*) \geq \epsilon\}} \{G(x) - I(x)\} < \Lambda$$

for every $\epsilon > 0$; if g is a bounded function on E which is continuous at x^* then

$$\lim_{\ell \to \infty} \int_E g(x) \tilde{K}_\ell[dx] = g(x^*).$$

§6 The Upper Bound in the General Setting

In this section we return to the programme, begun in §3, of exploring the consequences of the existence of the pressure in the thermodynamic limit.

Theorem 2

Let $\{K_\ell : \ell = 1, 2, \ldots\}$ be a sequence of probability measures on $[0, \infty)$ satisfying (P1) and (P2); then (LD3) holds with rate-function $I^\mu(\cdot)$ given by

$$I^\mu(x) = p(\mu) + f(x) - \mu x,$$

where $f(\cdot)$ is the free-energy, the Legendre transform of $p(\cdot)$: $f(x) = \sup\{\mu x - p(\mu)\}$.

Proof:

First consider an interval $I_1 = [0, \rho_1]$ with $\rho_1 < p'_-(\mu)$ (since $\mu \mapsto p(\mu)$ is convex, the left-hand derivative $p'_-(\mu)$ and the right-hand derivative $p'_+(\mu)$ exist for all μ in D). For each ℓ and each $\alpha < 0$, we have

$$K_\ell^\mu[I_1] = \int_{[0,\infty)} 1_{[0,\rho_1]}(x) K_\ell^\mu[dx] \leq \int_{[0,\infty)} e^{V_\ell \alpha(x-\rho_1)} K_\ell^\mu[dx]$$

$$= e^{V_\ell\{p_\ell(\mu+\alpha)-p_\ell(\mu)-\alpha\rho_1\}}. \tag{6.1}$$

Thus

$$\limsup_{\ell\to\infty} \frac{1}{V_\ell} \ln K_\ell^\mu[I_1] \leq p(\mu+\alpha) - p(\mu) - \alpha\rho_1 \, , \; \alpha < 0. \tag{6.2}$$

It follows that

$$\limsup_{\ell\to\infty} \frac{1}{V_\ell} \ln K_\ell^\mu[I_1] \leq \inf_{\alpha<0} \{p(\mu+\alpha) - p(\mu) - \alpha\rho_1\}$$

$$= -\{p(\mu) + \sup_{\alpha'<\mu} \{\alpha'\rho_1 - p(\alpha')\} - \mu\rho_1\}. \tag{6.3}$$

But

$$\sup_{\alpha<\mu}\{\alpha\rho_1 - p(\alpha)\} = \sup_{\alpha}\{\alpha\rho_1 - p(\alpha)\}$$

since $\rho < p'_-(\mu)$; hence

$$\limsup_{\ell\to\infty} \frac{1}{V_\ell} \ln K_\ell^\mu[I_1] \leq -I^\mu(\rho_1). \tag{6.4}$$

Next consider $I_2 = [\rho_2, \infty)$, where $\rho_2 > p'_+(\mu)$. It follows in analagous fashion that

$$\limsup_{\ell\to\infty} \frac{1}{V_\ell} \ln K_\ell^\mu[I_2] \leq -I^\mu(\rho_2). \tag{6.5}$$

Now let C be an arbitrary closed subset of $[0 \, , \, \infty)$; if $C \cap [p'_-(\mu), p'_+(\mu)]$ is non-empty then $\inf_C I^\mu(x) = 0$ and the inequality holds trivially since, for an arbitrary Borel set A, we have $K_\ell^\mu[A] \leq 1$; on the other hand, if $C \cap [p'_-(\mu), p'_+(\mu)]$ is empty, let (ρ_1, ρ_2) be the largest open interval containing $[p'_-(\mu), p'_+(\mu)]$ which does not intersect C so that $C \subset [0, \rho_1] \cup [\rho_2, \infty)$ and

$$\limsup_{\ell\to\infty} \frac{1}{V_\ell} \ln K_\ell^\mu[C] \le \limsup_{\ell\to\infty} \frac{1}{V_\ell} \ln \{K_\ell^\mu[I_1] + K_\ell^\mu[I_2]\}$$

$$\le (\limsup_{\ell\to\infty} \frac{1}{V_\ell} \ln K_\ell^\mu[I_1]) \vee (\limsup_{\ell\to\infty} \frac{1}{V_\ell} \ln K_\ell^\mu[I_2])$$

$$= (-I^\mu(\rho_1)) \vee (-I^\mu(\rho_2))$$

$$= -\inf_C I^\mu(x), \tag{6.6}$$

since $x \mapsto I^\mu(x)$ is decreasing on $[0, p'_-(\mu)]$ and increasing on $[p'_+(\mu), \infty)$

§7 The Berezin-Sinai Criterion for a First-Order Phase-Transition

We call an interval $[x_1, x_2]$ of the positive real axis a first-order phase-transition segment if the free-energy function $x \mapsto f(x)$ is linear for $x_1 \le x \le x_2$. Since $f(x) = \sup_\mu \{\mu x - p(\mu)\}$, each first-order phase-transition segment corresponds to a point μ at which the grand canonical pressure is non-differentiable: there exists μ such that

$$[x_1, x_2] = [p'_-(\mu), p'_+(\mu)].$$

On such an interval, the pressure as a function of the density is constant. In [7], Berezin and Sinai established a criterion for the existence of a first-order phase-transition segment; Dobrushin [8] simplified the proof considerably, pointing out that the criterion reduces the question of the existence of a phase-transition to the question of a "violation of the law of large numbers" in the grand canonical ensemble. Here we point out that the proof of the Berezin-Sinai criterion makes use only of the large-deviation upper bound and this, as we have seen, holds whenever the pressure exists.

The Berezin-Sinai Criterion

Suppose that, for some μ_o of μ, the rate function is symmetric about some point x_o:

$$I^{\mu_o}(x_o + y) = I^{\mu_o}(x_o - y)$$

for all y. Suppose also that for some $\delta > 0$:

$$P_\ell^{\mu_o}[|X_\ell - x_o| \ge \delta] \ge c > 0$$

for all ℓ sufficiently large. Then there is a first-order phase-transition at μ_o and the interval $[x_o - \delta, x_o + \delta]$ is contained in the phase-transition segment

$[p'_-(\mu_o), p'_+(\mu_o)]$.

<u>Proof</u>

Let $C = (-\infty, x_o - \delta] \cup [x_o + \delta, \infty)$; then, by hypothesis

$$\lim_{\sup} \frac{1}{V_\ell} \ln K_\ell^{\mu_o}[C] = 0; \tag{7.1}$$

by Theorem 2,

$$\limsup_{\ell \to \infty} \frac{1}{V_\ell} \ln K_\ell^{\mu_o}[C] \leq -\inf_C I^{\mu_o}(x) \leq 0. \tag{7.2}$$

Hence $\inf_C I^{\mu_o}(x) = 0$; by the symmetry of $I^{\mu_o}(\bullet)$ about x_o,

$$\inf\{I^{\mu_o}(x) : x \in [-\infty, x_o]\}$$

$$= \inf\{I^{\mu_o}(x) : x \in [x_o + \delta, \infty)\} = 0 \tag{7.3}$$

so that $x_o - \delta$ and $x_o + \delta$ must lie in $[p'_-(\mu_o), p'_+(\mu_o)]$ and therefore

$$[x_o - \delta, x_o + \delta] \subset [p'_-(\mu_o), p'_+(\mu_o)]$$

§8 <u>The Lower Bound in the General Setting</u>

We define the first-order phase-transition set F to be the union of the first-order phase-transition segments:

$$F = \cup_{\mu \in S}[p'_-(\mu), p'_+(\mu)]$$

where

$$S = \{\mu : p'_-(\mu) \neq p'_+(\mu)\}.$$

<u>Theorem</u> 3

<u>Let</u> $\{K_\ell : \ell = 1, 2, ...\}$ <u>be a sequence of probability measures on</u> $[0, \infty)$ <u>satisfying</u> (P1) <u>and</u> (P2); <u>let</u> G <u>be an open subset of</u> ran $\partial p \setminus F$ <u>where</u> $\partial p(\mu)$ <u>is the sub-differential of</u> p <u>at</u> μ; <u>then</u>

$$\liminf_{\ell \to \infty} \frac{1}{V_\ell} \ln K_\ell^{\mu}[G] \geq -\inf_G I^{\mu}(x).$$

Proof:

Let y be an arbitrary point of G; choose δ so that the neighbourhood $B_y^\delta = (y-\delta, y+\delta)$ is contained in G; then

$$K_\ell^\mu[G] \geq K_\ell^\mu[B_y^\delta] \int_{B_y^\delta} K_\ell^\mu[dx] = \int_{B_y^\delta} e^{-V_\ell c_\ell^\mu(x\,;\,\alpha)} K_\ell^{\mu+\alpha}[dx] \tag{8.1}$$

for all μ in the domain D of K_ℓ (by Lemma 2 of §3). Now choose α so that $y = p'(\mu+\alpha)$; this is possible because, by hypothesis, y is in ran $\partial p \setminus F$. Then

$$K_\ell^\mu[B_y^\delta] = e^{-V_\ell c_\ell^\mu(y\,;\,\alpha)} \int_{B_y^\delta} e^{-V_\ell \alpha(x-y)} K_\ell^{\mu+\alpha}[dx]$$

$$\geq e^{-V_\ell c_\ell^\mu(y\,;\,\alpha)}\, e^{-V_\ell \delta|\alpha|}\, K_\ell^{\mu+\alpha}[B_y^\delta]. \tag{8.2}$$

and, by Theorem 1, $\{K_\ell^{\mu+\alpha}\} \xrightarrow{w} \delta_y$ so that $K_\ell^{\mu+\alpha}[B_y^\delta] > \frac{1}{2}$ for all ℓ sufficiently large. Hence

$$\liminf_{\ell\to\infty} \frac{1}{V_\ell} \ln K_\ell^\mu[G] \geq -I^\mu_{(y)} - \delta|\alpha|;$$

but δ was an arbitrary positive number and y an arbitrary point of G; it follows that

$$\liminf_{\ell\to\infty} \frac{1}{V_\ell} \ln K_\ell^\mu[G] \geq \sup_G(-I^\mu(y)) = -\inf_G I^\mu(y) \tag{8.3}$$

§9 The Large Deviation Principle in the General Setting

In this section we put together the results of §6 and §8. First, we note some properties of the rate-function stemming from the convexity of the pressure $\mu \mapsto p(\mu)$. The free-energy $f(\bullet)$ is the Legendre transform of $p(\bullet)$:

$$f(x) = \sup_\mu\{\mu x - p(\mu)\}. \tag{9.1}$$

Hence

$$I^\mu(x) = p(\mu) + f(x) - \mu x \geq 0.$$

We may regard $I^\mu(\cdot)$ itself as the Legendre transform of the convex function $\alpha \mapsto p(\mu+\alpha) - p(\mu)$; it follows that $x \mapsto I^\mu(x)$ is a closed convex function and hence lower semi-continuous, so that (LD1) holds. Since $I^\mu(x) + p(\mu+\alpha) - p(\mu) - \alpha x \geq 0$, it follows that on the level set

$$L_m = \{x : I^\mu(x) \leq m\} \tag{9.2}$$

we have

$$\alpha x \leq m + p(\mu + \alpha) - p(\mu)\,; \tag{9.3}$$

hence, for $a > 0$

$$ax = \sup_{\alpha \in [-a,a]} \alpha x \leq m + \sup_{\alpha \in [-a,a]} p(\mu + \alpha) - p(\mu)$$

$$< \infty. \tag{9.4}$$

It follows that L_m is bounded; since $x \mapsto I^\mu(x)$ is lower semi-continuous, L_m is closed; hence L_m is compact and so (LD2) holds.

In §6 we saw that (LD3) holds whenever the pressure exists in the thermodynamic limit; on the other hand, it is clear from §8 that more is required for (LD4) to hold since it asserts the lower bound for all open sets while the existence of the pressure suffices to establish the lower bound only for open subsets of ran $\partial p \setminus F$. A sufficient condition for (LD4) to hold is that p exists and is differentiable on the whole of R and that ran $p' = [0, \infty)$; this is far from being necessary, however, as can be seen from the case of the free boson gas (see [3] in this volume). Nevertheless, this condition is satisfied sufficiently often to make the following theorem useful:

Theorem 4

Let $\{K_\ell : \ell = 1, 2, ..\}$ be a family of sequences of probability measures on $[0, \infty)$ defined for all values of μ in R and satisfying (P1) and (P2). Suppose that $p(\cdot)$ is differentiable and that ran $p' = [0, \infty)$; then, for each value of μ, the sequence $\{K^\mu_\ell : \ell = 1, 2, ..\}$ satisfies the large deviation principle with constants $\{V_\ell\}$ and rate-function $I^\mu(\cdot)$ given by $I^\mu(x) = p(\mu) + f(x) - \mu x$ where $f(x) = \sup_\mu \{\mu x - p(\mu)\}$.

§10 Remarks

I have attempted in this lecture to set out the results described in [1] in the framework established by Varadhan [2], thus showing the probabilistic consequences of the existence of the grand canonical pressure in the thermodynamic limit. Independently of [1], Ellis [9] proved a large deviation result for vector-valued random variables; his basic hypothesis is the existence of the limit of a sequence of cumulant generating functions, and Theorem 4 is a special case of his theorem.

REFERENCES

[1] J.T. Lewis and J.V. Pulè : The Equivalence of Ensembles in Statistical Mechanics, in Stochastic Analysis and its Applications, Proceedings, Swansea 1983 ed. A. Truman and D. Williams, LNM 1095, Springer : Heidelberg 1984.

[2] S.R.S. Varadhan : Asymptotic Probabilities and Differential Equations, Comm. Pure Apppl. Math. 19 261-286(1966).

[3] M. van den Berg, J.T. Lewis and J.V. Pulè : Large Deviations and the Boson Gas (this volume).

[4] L. van Hove : Quelques propriétés générales de l'intégrale de configuration d'un système de particules à une dimension, Physica 15 951-961 (1949).

[5] M. Schilder : Some asymptotic formulae for Wiener integrals, Trans. Amer. Math. Soc. 125 63-85 (1966).

[6] S.R.S. Varadhan : Large Deviations and Applications, Philadelphia : Society for Industrial and Applied Mathematics 1984.

[7] F.A. Berezin and Ya. G. Sinai : Existence of a Phase-Transition of the first kind in a lattice gas with attraction between particles, Trudy Mosk. Mat. Obshch 17, 197-212 (1967).

[8] R.L. Dobrushin : Existence of a phase-transition in models of a lattice gas, Proc. Fifth Berkeley Symposium, III, 73-87 (1967).

[9] R.S. Ellis : Large Deviations for a general class of random vectors, Ann. Probab. 12, 1-12 (1984).

Excursions from random boundaries

Paul McGill

School of Theoretical Physics, DIAS, 10 Burlington Road, Dublin 4, Ireland

The purpose of this short note is to show how the general theory of processes leads to a result on the decomposition of Markov processes at certain random times, including those times first considered by Williams [5]. The method, essentially due to Walsh [4], is to use a filtration indexed by the space variable. The results given here may hold for more general situations, but we have restricted our ambitions in the interests of clarity.

Let X_t be a non-singular Feller process taking values in R. For general results concerning such processes we shall rely mostly on [2]. Here non-singular means that at each starting point there is a non-zero probability of moving in either direction. When looking at the sample paths we shall, unless otherwise stated, consider only the cadlag version and we suppose that X_t is transient, either dying at a finite time or exiting to $+\infty$. Then introduce T_x as the first hitting time of the interval $(-\infty, x]$ and write L_x to represent the last leaving time of the same set. The usual conventions apply when either of these times is infinite, so that in particular we identify the sets $\{L_x = 0\}$ and $\{T_x = +\infty\}$. Notice how T_x is the terminal time associated to the coterminal time L_x in the sense of [2]. Also remark that the instead of the interval $(-\infty, x]$ we could consider the interval $(-\infty, x)$. This would give results of the same type provided hitting times are replaced by penetration times.

Definition For each real number x we write $\mathcal{X}_x^i = \sigma\{X_t : 0 \leq t \leq L_x\}$

It is clear that $\{\mathcal{X}_x^i, x \in R\}$ defines a filtration and we will suppose it to be complete. However it is not so obvious that it satisfies the usual conditions, and as can be seen from [4], this is crucial. We need some definitions. Using the results of [2] we find that the semi-group of X_t conditioned to converge to the interval

$(-\infty, x]$ before it dies, is given by

$$\tilde{\mathbf{E}}_y^x[f(X_t)] = \frac{\mathbf{E}_y[f(X_t); t < T_x < \infty]}{\mathbf{P}_y[T_x < \infty]} = \int \tilde{P}_t^x(y, dz) f(z)$$

Then the transition semigroup for the process $\{X_t \circ \theta_{L_x} : t > 0\}$, which is also strongly Markovian, is given by

$$\bar{\mathbf{E}}_y^x[f(X_t)] = \frac{\mathbf{E}_y[f(X_t); T_x = \infty]}{\mathbf{P}_y[T_x = \infty]} = \int \tilde{P}_t^x(y, dz) f(z) \frac{h(z, x)}{h(y, x)}$$

Here θ is the shift operator on the paths of X_t and we note how the formula is valid only for $t > 0$.

We shall make the following further assumptions regarding the semigroup. These are:

(1) if $h(x, y) = \mathbf{P}_x[T_y = +\infty]$ then for $y < x$ we have $\lim_{z \downarrow y} h(x, z) = h(x, y)$, and

(2) for each bounded continuous f the function $(x, y) \to \bar{\mathbf{E}}_y^x[f(X_t)]$ is bicontinuous on the closed upper diagonal $\{(x, y) : x \leq y\}$.

It is reasonably clear that our method cannot work in the absence of some continuity condition of this type.

Lemma 1 $\mathbf{P}[X_t \wedge X_{t-} = x, t > L_x] = 0$

Proof: This follows immediately from the continuity assumption (1) since by the above formula if $y > x$ then

$$\bar{\mathbf{P}}_y^x[T_{x+\varepsilon} < +\infty] = \frac{\mathbf{P}_y[T_{x+\varepsilon} < +\infty; T_x = \infty]}{\mathbf{P}_y[T_x = \infty]} = 1 - \frac{\mathbf{P}_y[T_{x+\varepsilon} = +\infty]}{\mathbf{P}_y[T_x = +\infty]}$$

which decreases to zero with ε.

Lemma 2 Almost surely $L_{x+\frac{1}{n}}$ converges to L_x.

Proof: We let $\bar{T}_{x+\frac{1}{n}}$ be the hitting time $\inf\{t > L_x : X_t \geq x + \frac{1}{n}\}$. Then by the strong Markov property [2] of $X_t \circ \theta_{L_x}$ at the time $\bar{T}_{x+\frac{1}{n}}$ and Lemma 1, we see that $L_{x+\delta} \leq \bar{T}_{x+\frac{1}{n}}$ when δ (random) is sufficiently small. However by right continuity of the paths $\lim \bar{T}_{x+\frac{1}{n}} = L_x$ and we have the result.

Lemma 3 The filtration $\{\mathcal{X}_x^f, x \in R\}$ is right continuous.

Proof: It suffices to show that for all bounded measurable functions A of the process $Y_t = X_t \circ \theta_{L_{xm}}$ we have almost sure equality

$$\mathbf{E}[A|\mathcal{X}_x^i] = \lim \mathbf{E}[A|\mathcal{X}_{x_n}^i]$$

where $x_n \downarrow x$. So fix $n > m$ and suppose that A is bounded, of the form $f_1(Y_{t_1})f_2(Y_{t_2})\dots f_k(Y_{t_k}) \circ \theta(L_{x_m})$ with $0 < t_1 < t_2 < \dots < t_k$. By a density argument, using Lemma 2 and the right continuity X_t, this suffices. Now use the strong Markov property at time L_{x_n} to get

$$\begin{aligned}\mathbf{E}[f_1(Y_{t_1})f_2(Y_{t_2})\dots f_k(Y_{t_k}) \circ \theta(L_{x_m})|\mathcal{X}_{x_n}^i] &= \\ \bar{\mathbf{E}}^{x_n}_{X(L(x_n))}[f_1(Y_{t_1})f_2(Y_{t_2})\dots f_k(Y_{t_k}) \circ \theta(L_{x_m})].\end{aligned}$$

and we take the limit in n. By the previous lemma, assumption (2), and the strong Markov property of $X_t \circ \theta(L_x)$ we see that this converges to $\mathbf{E}[f_1(Y_{t_1})f_2(Y_{t_2})\dots$ $\dots f_k(Y_{t_k}) \circ \theta(L_{x_m})|\mathcal{X}_x^i]$. This completes the proof.

Since X_t is a transient process we know that it has a minimum limiting value ξ. By the strong Markov property, and also since X_t is non-singular, it follows that if ξ is attained then this happens at most once. The interesting stopping times Z of the filtration $\mathcal{X}_x^i$ are those which satisfy the condition $Z \geq \xi$.

Theorem 1 (1) The minimum ξ is an $\mathcal{X}_x^i$ stopping time.
(2) If Z is any $\mathcal{X}_x^i$ stopping time then for every bounded measurable function $\mathcal{F}$of the path

$$\mathbf{E}[\mathcal{F} \circ \theta(L_x) 1_{(Z\leq x)}|\mathcal{X}_Z^i] = \bar{\mathbf{E}}^x_{X(L_x)}[\mathcal{F}]1_{(Z\leq x)}$$

Proof: (1) By remarks made above concerning the minimum we have the almost sure equality $\{\xi < x\} = \{T_x < +\infty\}$. Because the rhs is an element of $\mathcal{X}_x^i$ the proof follows, using right continuity of the filtration.
(2) This follows by the strong Markov property applied at the time L_x to X_t, using the fact that by definition $A1_{(Z\leq x)}$ is $\mathcal{X}_x^i$ measurable for each $A \in \mathcal{X}_Z^i$.

One might also be interested, not just in the final excursion from $(-\infty, x]$, but also in the initial excursion. To look at this we define another filtration; the analogue of $\mathcal{X}_x^i$ in reversed time.

Definition We shall write $\mathcal{X}_x^f = \sigma\{X(T_x-), X_t \circ \theta_{T_x} : t \geq 0\}$.

We are now forced into making the assumption that X_t in reversed time satisfies the conditions imposed on the semigroup of X_t. This leads to a little imprecision.

For example if the death time is infinite then it suffices to suppose that we reverse from L_z for some sufficiently large z. Note that by [2] this is again a death time for X_t, and that the killed process retains the Feller property. Also, there is a slight difficulty about the status of X_0 which can be resolved in the usual way by randomising independently and then conditioning at the end. We will write $\mathcal{X}_x = \mathcal{X}_x^f \cap \mathcal{X}_x^i$, and $\hat{\mathbf{E}}_y^x$ shall be the reversed time counterpart of $\bar{\mathbf{E}}_y^x$.
It is clear, under these assumptions, that all three filtrations defined above are right continuous and complete. Moreover we have the following extension of Theorem 1, where we use k_t for the killing operator.

Theorem 2 Let $\mathcal{F}$ and $\mathcal{G}$ be two bounded continuous functionals of the process X_t.

(1) The minimum ξ is an $\mathcal{X}_x^f$ stopping time.
(2) If Z is any $\mathcal{X}_x^f$ stopping time then

$$\mathbf{E}[\mathcal{G} \wedge k(T_x-)1_{(Z\le x)}|\mathcal{X}_Z^f] = \hat{\mathbf{E}}^x_{X(T_x-)}[\mathcal{G}]1_{(Z\le x)}$$

(3) If Z is any $\mathcal{X}_x$ stopping time then

$$\mathbf{E}[\mathcal{G} \wedge k(T_x-)\mathcal{F} \circ \theta(L_x)1_{(Z\le x)}|\mathcal{X}_Z] = \hat{\mathbf{E}}^x_{X(T_x-)}[\mathcal{G}]\bar{\mathbf{E}}^x_{X(L_x)}[\mathcal{F}]1_{(Z\le x)}$$

It is this last part which is most interesting from our point of view, for we have the following immediate consequence.

Corollary A (1) If Z is any $\mathcal{X}_x^i$ stopping time we have

$$\mathbf{E}[\mathcal{F} \circ \theta(L_Z)|\mathcal{X}_Z^i] = \bar{\mathbf{E}}^Z_{X(L_Z)}[\mathcal{F}]$$

(2) If Z is any $\mathcal{X}_x^f$ stopping time then

$$\mathbf{E}[\mathcal{G} \wedge k(T_Z-)|\mathcal{X}_Z^f] = \hat{\mathbf{E}}^Z_{X(T_Z-)}[\mathcal{G}]$$

(3) If Z is any $\mathcal{X}_x$ stopping time then

$$\mathbf{E}[\mathcal{G} \wedge k(T_Z-)\mathcal{F} \circ \theta(L_Z)|\mathcal{X}_Z] = \hat{\mathbf{E}}^Z_{X(T_Z-)}[\mathcal{G}]\bar{\mathbf{E}}^Z_{X(L_Z)}[\mathcal{F}]$$

Proof: We only look at (1). This is obtained by replacing x by Z in Theorem 1 (2); which is clear if Z is a simple stopping time since Z is $\mathcal{X}_Z^i$ measurable. For the general case we take a sequence of simple $\mathcal{X}_Z^f$ measurable random times $Z_n \downarrow Z$, noting that all these must be stopping times, and then take the limit using Lemma 2 and the assumption of bicontinuity of the semi-group.

Note how (3) above says that the initial and final excursions to Z are conditionally independent given the σ-field $\mathcal{X}_Z$, with respective semigroups $\hat{\mathbf{E}}^Z_{X(T_Z-)}$ and $\bar{\mathbf{E}}^Z_{X(L_Z)}$. This result can undoubtedly be found in a more general form in [3], but its statement is so natural and clearcut that I think it worth noting separately. One can compare the proof given here with those of [1],[3] and [5]. We also highlight the following well-known particular case.

Corollary B The initial and final excursions of X_t to the minimum ξ are conditionally independent given ξ, with respective semi-groups $\hat{\mathbf{E}}^\xi$ and $\bar{\mathbf{E}}^\xi$.

Proof: We only need to show that $\mathcal{X}_\xi$ is the σ-field generated by ξ. For this we use the right continuity of X_t, its non-singularity, and the strong Markov property.

Remark that if the process is continuous then $\hat{\mathbf{E}}$ is indeed the the time reversal of $\tilde{\mathbf{E}}$, so this is the appropriate generalisation of the Williams decomposition [5]. We offer also the following comments on how this note relates to the results in [3]. There the times used are more general, and the assumptions on the process are weaker and more probabilistic. However we think our method of applying Walsh's idea to the decomposition problem is interesting, and conceptually simpler. Moreover it highlights Millar's comment that we are studying the process 'after last exit from a boundary (suitably) chosen at random'.

REFERENCES

1. **P. Greenwood and J. Pitman** *Fluctuation identities for Lévy processes and splitting at the maximum. Adv. App. Prob.* **12** 893-902 (1980).

2. **P.A. Meyer, R.T. Smythe and J.B. Walsh** *Birth and death of Markov processes. Proc. 6th Berkeley Symp., Vol. III*, 295-306 (1972).

3. **P.W. Millar** *Random times and decomposition theorems. Proc. Symp. in Pure Mathematics, Vol. 31*, 91-103 (1977).

4. **J.B. Walsh** *Excursions and local time. In: Temps Locaux, Astérisque Vol. 52/3,* Soc. Math. France, 1977.

5. **D. Williams** *Decomposing the Brownian path. Bull. Amer. Math. Soc.* **76**, 871-3 (1970).

A SOLUTION OF THE INTEGRAL EQUATION

$$\int_0^\infty (\sin(\tfrac{\pi}{4}+\theta)e^{-\theta y} + \sin(\tfrac{\pi}{4}-\theta)e^{\theta y})\Pi(x,y)dy = \sqrt{2}\cosh\theta x$$

IN CONVOLUTION FORM AND A PROBLEM IN DIFFUSION THEORY

by

Malcolm T. McGregor
Department of Mathematics and Computer Science
University College of Swansea
Singleton Park, Swansea SA2 8PP

1. Introduction

Let $B_x(s)$ be a reflecting Brownian motion on $(0,\infty)$ with $B_x(0) = x$. Let τ^+ be the first time s that the sojourn time of $(1,\infty)$ for B_x up to time s exceeds the sojourn time of $[0,1]$ up to time s, and define $Y^+ = B_x(\tau^+)$. Then in [1] it was established that for $0 < x < 1$ the probability density of Y^+ is $\Pi(x,y)$ in the sense that $\mathbb{P}^x(Y^+ \in (1+y, 1+y+dy)) = \Pi(x,y)dy$, where Π satisfies

$$\int_0^\infty (\cosh\theta\cos\theta y + \sinh\theta\sin\theta y)\Pi(x,y)dy = \cosh\theta x, \qquad \theta > 0. \tag{1}$$

For further discussion of this remarkable result, additional references and some generalisations see [2] and [3].

In [1] amazingly the closed form solution to the above problem is obtained by ad hoc methods in the form

$$\Pi(x,y) = \frac{\cosh\frac{1}{2}\pi y(\sinh\frac{1}{2}\pi y\cos\frac{1}{2}\pi x)^{\frac{1}{2}}}{\sqrt{2}(\sinh^2\frac{1}{2}\pi y + \cos^2\frac{1}{2}\pi x)}. \tag{2}$$

Here we give the first published constructive proof of this result. We challenge the reader to generalise our constructive result to include the more general class of problems discussed in [3]. For an equally intriguing set of results see the forthcoming paper by Paul McGill.

We begin by considering the integral equation

$$\int_0^\infty (\sin(\tfrac{\pi}{4}+\theta)e^{-\theta y} + \sin(\tfrac{\pi}{4}-\theta)e^{\theta y})\Pi(x,y)dy = \sqrt{2}\cosh\theta x \tag{3}$$

where x and θ are complex, and we show by Laplace transform methods that it has a solution in convolution form

$$\Pi(x,y) = \sqrt{(2\pi)}\int_0^y \frac{G(x,y-\nu)d\nu}{\sqrt{(\sinh\frac{1}{2}\pi\nu)}}, \tag{4}$$

where

$$G(x,\tau) = G(x,-\tau) = \frac{\sqrt{\pi}}{16}\{[\cosh\tfrac{1}{2}\pi(x+\tau)]^{-3/2} + [\cosh\tfrac{1}{2}\pi(x-\tau)]^{-3/2}\}. \tag{5}$$

Furthermore, by setting $a = e^{\frac{1}{2}\pi y}$, $b = -e^{-\frac{1}{2}\pi y}$, $p = e^{\frac{1}{2}\pi x}$ and $q = e^{-\frac{1}{2}\pi x}$ in

$$\int\frac{d\nu}{(a\nu+b)^{\frac{1}{2}}(p\nu+q)^{3/2}} = \frac{2}{(aq-bp)}\left(\frac{a\nu+b}{p\nu+q}\right)^{\frac{1}{2}} \tag{6}$$

we show that $\Pi(x,y)$ as given by (4) and (5) may be written as

$$\Pi(x,y) = \frac{\cosh\frac{1}{2}\pi y(\sinh\frac{1}{2}\pi y\cosh\frac{1}{2}\pi x)^{\frac{1}{2}}}{\sqrt{2}(\sinh^2\frac{1}{2}\pi y+\cosh^2\frac{1}{2}\pi x)}. \tag{7}$$

Equation (1) is the special case of (3) obtained by replacing θ and x by $i\theta$ and ix respectively. Thus we obtain (2) as a solution to (1) by taking ix in place of x.

2. A key identity

Let x and θ be complex with $-1 < \operatorname{im} x < 1$ and $-\frac{\pi}{4} < \operatorname{re}\theta < \frac{\pi}{4}$; the conditions on x and θ are sufficient to guarantee that the various integrals exist and for the gamma and beta functions to be defined. With $G(x,\tau)$ given by (5), we shall find it convenient to set $\tilde{G}(x,\theta) = \mathcal{L}[G(x,\tau)](\theta)$, and we can then proceed to establish the identity

$$\tilde{G}(x,\theta) + \tilde{G}(x,-\theta) = (\pi\sqrt{2})^{-1}\Gamma(\tfrac{3}{4}+\tfrac{\theta}{\pi})\Gamma(\tfrac{3}{4}-\tfrac{\theta}{\pi})\cosh\theta x. \tag{8}$$

We begin by showing that

$$\tilde{G}(x,\theta) + \tilde{G}(x,-\theta) = \frac{\sqrt{\pi}}{8}\int_{-\infty}^\infty \frac{\cosh\theta\tau\, d\tau}{[\cosh\frac{1}{2}\pi(x+\tau)]^{3/2}}. \tag{9}$$

Clearly,

$$\tilde{G}(x,\theta) + \tilde{G}(x,-\theta) = 2\int_0^\infty \cosh\theta\tau\, G(x,\tau)d\tau$$

and substituting for $G(x,\tau)$ from (5) readily produces (9). To complete the proof of (8) we show that the right-hand sides of (8) and (9) are equal. We begin with the well-known beta function formula

$$B(m,n) = \int_0^\infty \frac{v^{m-1}dv}{(1+v)^{m+n}} \qquad (\mathrm{re}\, m > 0,\ \mathrm{re}\, n > 0)$$

and we set $e^{2u} = v$ to get, with $m = \frac{3}{4} + s$ and $n = \frac{3}{4} - s$,

$$\int_{-\infty}^\infty \frac{e^{2su}\, du}{(\cosh u)^{3/2}} = \sqrt{2}\, B(\tfrac{3}{4} + s, \tfrac{3}{4} - s).$$

With $u = x + t$, we deduce that

$$\int_{-\infty}^\infty \frac{e^{2st}\, dt}{[\cosh(x+t)]^{3/2}} = \sqrt{2}\, e^{-2sx} B(\tfrac{3}{4} + s, \tfrac{3}{4} - s),$$

and with s replaced by $-s$, we get on adding the two formulae

$$\int_{-\infty}^\infty \frac{\cosh 2st\, dt}{[\cosh(x+t)]^{3/2}} = \sqrt{2}\cosh 2sx\, B(\tfrac{3}{4} + s, \tfrac{3}{4} - s).$$

With x replaced by $\frac{1}{2}\pi x$, $t = \frac{1}{2}\pi\tau$, $s = \frac{\theta}{\pi}$ we only need some well-known properties of the beta and gamma functions to complete the proof of (8).

3. A Wiener-Hopf identity

Let $\tilde{\Pi}(x,\theta) = \mathcal{L}[\Pi(x,y)](\theta)$, with $\Pi(x,y)$ as given by (4) and (5). Our integral equation (3) then takes the form

$$\sin(\tfrac{\pi}{4} + \theta)\tilde{\Pi}(x,\theta) + \sin(\tfrac{\pi}{4} - \theta)\tilde{\Pi}(x,-\theta) = \sqrt{2}\cosh\theta x. \tag{10}$$

We proceed to show that this Wiener-Hopf identity is satisfied by

$$\tilde{\Pi}(x,\theta) = \frac{2}{\sqrt{\pi}}\tilde{G}(x,\theta)\, B(\tfrac{1}{4} + \tfrac{\theta}{\pi}, \tfrac{1}{2}). \tag{11}$$

Substituting into the left-hand side of (10) for $\tilde{\Pi}(x,\theta)$ and $\tilde{\Pi}(x,-\theta)$ from (11), we have

$$\frac{2}{\sqrt{\pi}}\left(\sin(\tfrac{\pi}{4}+\theta)\tilde{G}(x,\theta)B(\tfrac{1}{4}+\tfrac{\theta}{\pi},\tfrac{1}{2}) + \sin(\tfrac{\pi}{4}-\theta)\tilde{G}(x,-\theta)B(\tfrac{1}{4}-\tfrac{\theta}{\pi},\tfrac{1}{2})\right)$$

$$= \frac{2\Gamma(\frac{1}{2})}{\sqrt{\pi}}\left\{\frac{\sin(\frac{\pi}{4}+\theta)\Gamma(\frac{1}{4}+\frac{\theta}{\pi})\tilde{G}(x,\theta)}{\Gamma(\frac{3}{4}+\frac{\theta}{\pi})} + \frac{\sin(\frac{\pi}{4}-\theta)\Gamma(\frac{1}{4}-\frac{\theta}{\pi})\tilde{G}(x,-\theta)}{\Gamma(\frac{3}{4}-\frac{\theta}{\pi})}\right\}.$$

With $z = \frac{1}{4}+\frac{\theta}{\pi}$ and $z = \frac{1}{4}-\frac{\theta}{\pi}$ in $\Gamma(z)\Gamma(1-z) = \pi/\sin \pi z$ the left-hand side of (10) may be reduced to

$$\frac{2\pi(\tilde{G}(x,\theta) + \tilde{G}(x,-\theta))}{\Gamma(\frac{3}{4}+\frac{\theta}{\pi})\Gamma(\frac{3}{4}-\frac{\theta}{\pi})}$$

since $\Gamma(\frac{1}{2}) = \sqrt{\pi}$. Using (8) this becomes $\sqrt{2}\cosh\theta x$ and we have shown that (10) is satisfied by $\tilde{\Pi}(x,\theta)$ as given in (11). To establish (11) we apply the Laplace transform to the convolution (4) to get

$$\tilde{\Pi}(x,\theta) = \sqrt{(2\pi)}\tilde{G}(x,\theta)\mathcal{L}[(\sinh\tfrac{1}{2}\pi\nu)^{-\frac{1}{2}}](\theta)$$

and it only remains to prove that

$$\mathcal{L}[(\sinh\tfrac{1}{2}\pi\nu)^{-\frac{1}{2}}](\theta) = \sqrt{2}\,\pi^{-1}B(\tfrac{1}{4}+\tfrac{\theta}{\pi},\tfrac{1}{2}). \tag{12}$$

To obtain (12) we simply set $e^{-\pi\nu} = u$ in the Laplace integral to get

$$\frac{\sqrt{2}}{\pi}\int_0^1 u^{\frac{\theta}{\pi}-\frac{3}{4}}(1-u)^{-\frac{1}{2}}du$$

which is the required beta function integral. It is now clear that the convolution (4) satisfies our integral equation (3).

4. Evaluation of the convolution

To evaluate the convolution (4) we set $y-\nu = -\tau$ to get

$$u(x,y) = -\sqrt{(2\pi)}\int_0^{-y}\frac{G(x,\tau)\,d\tau}{[\sinh\frac{1}{2}\pi(y+\tau)]^{\frac{1}{2}}}, \tag{13}$$

and if we substitute in for $G(x,\tau)$ from (5) then we have two integrals to determine. Let $X = \frac{1}{2}\pi x$, $Y = \frac{1}{2}\pi y$ and consider

$$I(X,Y) = \int_0^{-Y}\frac{dt}{[\sinh(Y+t)]^{\frac{1}{2}}[\cosh(X+t)]^{3/2}};$$

with $e^{2t} = \nu$ this becomes

$$\tfrac{1}{2}I(X,Y) = \int_1^{e^{-2Y}} \frac{d\nu}{(e^{Y}\nu - e^{-Y})^{\frac{1}{2}}(e^{X}\nu + e^{-X})^{3/2}}$$

and (6) with $a = e^{Y}$, $b = -e^{-Y}$, $p = e^{X}$ and $q = e^{-X}$ gives

$$\tfrac{1}{2}I(X,Y) = \frac{-2}{(e^{Y-X} + e^{X-Y})}\left(\frac{e^{Y} - e^{-Y}}{e^{X} + e^{-X}}\right)^{\frac{1}{2}}.$$

Replacing X by -X and adding gives

$$\tfrac{1}{2}(I(X,Y) + I(-X,Y)) = \frac{-2\cosh Y\cosh X}{(\sinh^{2}Y + \cosh^{2}X)}\left(\frac{\sinh Y}{\cosh X}\right)^{\frac{1}{2}},$$

and setting $\tau = \frac{2t}{\pi}$ in our integral (13) gives

$$\Pi(x,y) = \frac{-1}{4\sqrt{2}}\{I(\tfrac{1}{2}\pi x, \tfrac{1}{2}\pi y) + I(-\tfrac{1}{2}\pi x, \tfrac{1}{2}\pi y)\}$$

which is (7).

REFERENCES

[1] R.R. LONDON, H.P. McKEAN, L.C.G. ROGERS and DAVID WILLIAMS, A martingale approach to some Wiener-Hopf problems, II, Séminaire de Probabilités XVI, Springer Lecture Notes in Math. 920, 68-90.

[2] R.R. LONDON, H.P. McKEAN, L.C.G. ROGERS and DAVID WILLIAMS, A martingale approach to some Wiener-Hopf problems, I, Séminaire de Probabilités XVI, Springer Lecture Notes in Math. 920, 41-67.

[3] N. BAKER, Some integral equalities in Wiener-Hopf theory, Stochastic analysis and applications, Springer Lecture Notes in Math. 1095, 169-186.

FORMULÆ FOR THE HEAT KERNEL OF AN ELLIPTIC OPERATOR EXHIBITING SMALL-TIME ASYMPTOTICS

by

Keith D. Watling
Department of Mathematics and Computer Science,
University College of Swansea,
Singleton Park,
Swansea. SA2 8PP
Great Britain

§0 Introduction

Let L be a smooth scalar second-order elliptic differential operator on a smooth connected n-dimensional manifold M. The inverse of the symbol of this operator is the natural choice of smooth Riemannian structure on M. The 'Levi-Civita' connection of this natural Riemannian structure decomposes the operator as $L = \frac{1}{2}\Delta + b + c$, where Δ is the 'Laplace-Beltrami' operator for this natural Riemannian structure, b is a smooth vector field and c is a smooth real-valued function on M.

Let $p(t,x,y)$ be the fundamental solution of the heat equation,

$$\frac{\partial f}{\partial t}(t,x) = Lf_t(x),$$

with respect to the natural Riemannian measure.

We wish to examine the small-time behaviour of $p(t,x,y)$. This has been extensively studied by numerous authors, see for example Molchanov [1], Kifer [1], Azencott et al [1], Azencott [1], Bismut [1], Ikeda and Watanabe [1] and references contained in these. We will give an extension of the simple ideas of Elworthy and Truman [1] and Elworthy [1] (Chapter IX §12), where the emphasis (motivated by consideration of the Schrödinger equation) is on obtaining exact formulæ for $p(t,x,y)$ which do not lose any information: for example involve all the geodesics not just any minimal one, which is all that is relevant in the small-time asymptotic expansion approximations to $p(t,x,y)$. See also Arede [1], Elworthy, Ndumu and Truman [1], Ndumu [1], [2] and Watling [1] for other extensions of these ideas.

Under the simplifying assumption that y is a pole of the natural Riemannian manifold we obtain exact expressions for $p(t,x,y)$ which clearly exhibit the small-time asymptotic behaviour to any order. These may be extended to the case that the exponential map is a covering map in a similar manner to Elworthy [1] (Chapter

IX §12 Remark 12D(i)) and Arede [1]: so in particular we can obtain such formulæ for 'Cartan-Hadamard' manifolds and connected nilpotent Lie groups. Moreover the terms in the exact expressions will still give the correct terms in the small-time asymptotic expansion approximations to $p(t,x,y)$ on more general complete manifolds: where there is a unique minimal geodesic between x and y along which they are not conjugate (in other words x is not on the cut locus of y), because by Varadhan's estimate (see Azencott et al [1]) the asymptotic behaviour only depends on the behaviour in a neighbourhood of this geodesic.

§1 Some Riemannian Geometry

(1.1) Definitions.

We assume that y is a pole of the natural Riemannian manifold i.e., the exponential map based at y is a diffeomorphism, so there is a unique geodesic parameterised to take unit time between y and any point of the manifold. So we may define the following functions on M motivated by the discussion in §6 of Watling [1]:

$\Theta_y(x)$ = the *square root* of the Jacobian determinant with respect to the natural Riemannian structure of the inverse of the exponential map based at y, at the point x, i.e., the square root of the inverse of Ruse's invariant, see Besse [1] and Elworthy and Truman [1]. (1.1.1)

$$B_y(x) = \exp\left\{\int_0^1 \langle \gamma'(u), b(\gamma(u))\rangle \, du\right\}, \tag{1.1.2}$$

where γ is the unique geodesic from x to y parameterised to take unit time and γ' is its velocity, (the integrand is the work done by b in moving along the geodesic γ),

$$C_y(x) = \Theta_y(x) B_y(x), \tag{1.1.3}$$

$$E_y(x) = \frac{1}{2} d(x,y)^2, \tag{1.1.4}$$

where $d(x,y)$ is the natural Riemannian distance between x and y, ($E_y(x)$ is the energy of the geodesic γ). Finally we may define the following functions on $\mathbf{R}^+ \times M$:

$$r_y(t,x) = \frac{1}{\sqrt{(2\pi t)^n}} \Theta_y(x) \exp\left\{\frac{-E_y(x)}{t}\right\}, \tag{1.1.5}$$

$$q_y(t,x) = B_y(x) r_y(t,x). \tag{1.1.6}$$

(1.2) **Remark.** We need the important fact from Elworthy [1] (Chapter IX §12 Example 12D) that if $f(x)$ is a real valued function on M which depends only on $r = d(x,y)$ (i.e., it is invariant under rotations about y) then:

$$\Delta f(x) = \frac{d^2 f}{dr^2}(r) + \left\{\frac{n-1}{r} - 2\frac{\partial}{\partial r} \log \Theta_y(x)\right\} \frac{df}{dr}(r). \tag{1.2.1}$$

(1.3) **Lemma.**

$$\frac{\partial}{\partial t} r_y(t,x) = \frac{1}{2}\Delta r_y(t,x) - \frac{\Delta\Theta_y(x)}{2\Theta_y(x)} r_y(t,x).$$

Proof: Using the fact that $\Delta \exp(f) = \left\{\Delta f + \|\nabla f\|^2\right\} \exp(f)$, for real valued functions f on M, we see that:

$$\begin{aligned}
\frac{1}{2}\Delta r_y(t,x) &= \frac{1}{2}\left\{\Delta\left(\log\Theta_y - \frac{1}{t}E_y\right)(x) + \left\|\nabla \log\Theta_y(x) - \frac{1}{t}\nabla E_y(x)\right\|^2\right\} r_y(t,x), \\
&= \frac{1}{2}\left\{\Delta \log\Theta_y(x) - \frac{1}{t}\Delta E_y(x) - 2\left\langle \nabla \log\Theta_y(x), \nabla E_y(x)\right\rangle\right. \\
&\qquad \left. + \|\nabla \log\Theta_y(x)\|^2 + \frac{1}{t^2}\|\nabla E_y(x)\|^2\right\} r_y(t,x).
\end{aligned}$$

So from (1.2.1) applied to E_y and using the facts that

$$\Delta f = \left\{\Delta \log f + \|\nabla \log f\|^2\right\} f,$$

for real valued functions f on M, and

$$\|\nabla E_y(x)\|^2 = 2E_y(x)$$

we deduce that:

$$\begin{aligned}
\frac{1}{2}\Delta r_y(t,x) &= \left\{\frac{\Delta\Theta_y(x)}{2\Theta_y(x)} - \frac{n}{2t} + \frac{1}{t^2}E_y(x)\right\} r_y(t,x) \\
&= \frac{\Delta\Theta_y(x)}{2\Theta_y(x)} r_y(t,x) + \frac{\partial r_y}{\partial t}(t,x)
\end{aligned}$$

(1.4) **Proposition.** *If $p_y(t,x,z)$ is the fundamental solution of the heat equation:*

$$\frac{\partial f}{\partial t}(t,x) = Lf_t(x) - \frac{LC_y(x)}{C_y(x)} f_t(x),$$

with respect to the natural Riemannian measure. Then $p_y(t,x,y) = q_y(t,x)$.

Proof: As in Elworthy and Truman [1] , Elworthy [1] (Chapter IX §12B), we see that $q_y(t,x)$ (considered as a distribution with respect to the natural Riemannian measure) tends to the delta function at y as t tends to 0, by a simple change of variable argument.

By (1.3) and the fact that $\Delta(fg) = f\Delta g + g\Delta f + 2\langle\nabla f, \nabla g\rangle$, for real valued functions f and g:

$$\begin{aligned}\frac{\partial q_y}{\partial t}(t,x) &= B_y(x)\left\{\frac{1}{2}\Delta r_y(t,x) - \frac{\Delta\Theta_y(x)}{2\Theta_y(x)} r_y(t,x)\right\},\\ &= \frac{1}{2}\Delta q_y(t,x) + \left\{\frac{1}{t}\langle\nabla\log B_y(x), \nabla E_y(x)\rangle - \frac{\Delta C_y(x)}{2C_y(x)}\right\} q_y(t,x),\\ &= Lq_y(t,x) + \left\{\frac{1}{2}\langle b(x) + \nabla\log B_y(x), \nabla E_y(x)\rangle - \frac{LC_y(x)}{C_y(x)}\right\} q_y(t,x).\end{aligned}$$

Observe that if b is a gradient then $b + \nabla\log B_y = 0$, so in this case we have the result required. (1.4.1)

In general we will prove that

$$\langle b(x) + \nabla\log B_y(x), \nabla E_y(x)\rangle = 0. \tag{1.4.2}$$

i.e., $b + \nabla\log B_y$ only rotates about y.

Firstly note that we have $\nabla E_y(x) = \gamma'(0)$, where γ is the unique geodesic from x to y parameterised to take unit time, so:

$$\langle\nabla\log B_y(x), \nabla E_y(x)\rangle = \frac{d}{dr}\log B_y(\gamma(r))\bigg|_{r=0}$$

But:

$$\log B_y(\gamma(r)) = \int_0^1 \langle\alpha'(u), b(\alpha(u))\rangle\, du,$$

where α is the unique geodesic from $\gamma(r)$ to y parameterised to take unit time,

$$= \int_r^1 \langle(\gamma'(s), b(\gamma(s))\rangle\, ds,$$

where γ is as above, so:

$$\frac{d}{dr}\log B_y(\gamma(r))\bigg|_{r=0} = -\langle\gamma'(0), b(\gamma(0))\rangle = -\langle\nabla E_y(x), b(x)\rangle,$$

whence the result.

§2 Semi-Classical Semigroups and Bridge Processes

(2.1) **Definition.** Let $P(t)$ be the semigroup associated with $p(t,x,y)$. Define the operators $Q_y(t,s)$ by:

$$\{Q_y(t,s)f\}(x) = q_y(t,x)^{-1}\{P(t-s)(q_y(s,.)f(.))\}(x), \tag{2.1.1}$$

for $t \geq s > 0$, on smooth functions of compact support.

(2.2) **Remark.**

Formally these operators form a two parameter semigroup i.e.,

$$Q_y(t,s)Q_y(s,r) = Q_y(t,r).$$

This is like a 'Doob h-transform' of the semigroup $P(t)$ but with respect to a function q_y on space-time that is only 'approximately harmonic' for the operator: $\frac{\partial}{\partial t} - L$, with 'error' given by (1.4).

(2.3) **Lemma.** *For f smooth of compact support we have:*

$$\frac{\partial}{\partial s}\{Q_y(t,t-s)f\}(x) = \{Q_y(t,t-s)(I_y(t-s)f)\}(x), \text{for } t > s \geq 0, \quad (2.3.1)$$

and

$$\frac{\partial}{\partial t}\{Q_y(t,s)f\}(x) = \{I_y(t)(Q_y(t,s)f)\}(x), \text{for } t \geq s > 0, \quad (2.3.2)$$

where

$$\begin{aligned}\{I_y(t)f\}(z) &= \{(L-c)f\}(z) + \langle(\nabla \log q_y(t,z), \nabla f(z)\rangle + \frac{LC_y(z)}{C_y(z)}f(z), \\ &= \frac{L(C_y f)(z)}{Cy(z)} - \frac{1}{t}\langle \nabla E_y(z), \nabla f(z)\rangle \quad (2.3.3) \\ &= \frac{1}{2}\Delta f(z) + \left\langle \nabla \log \Theta_y(z) + (b + \nabla \log B_y)(z) - \frac{1}{t}\nabla E_y(z), \nabla f(z)\right\rangle \\ &+ \left\{\frac{\Delta \Theta_y(z)}{2\Theta_y(z)} + \frac{1}{2}\Delta \log B_y(z) + \langle \nabla \log \Theta_y(z), (b + \nabla \log B_y)(z)\rangle \right. \\ &\left. + \frac{1}{2}\|\nabla \log B_y(z)\|^2 + \langle b(z), \nabla \log B_y(z)\rangle + c(z)\right\} f(z) \quad (2.3.4)\end{aligned}$$

In particular when b is a gradient we have:

$$\begin{aligned}\{I_y(t)f\}(z) = \frac{1}{2}\Delta f(z) &+ \langle \nabla \log r_y(t,z), \nabla f(z)\rangle \\ &+ \left\{\frac{\Delta \Theta_y(z)}{2\Theta_y(z)} - \frac{1}{2}\operatorname{div} b(z) - \frac{1}{2}\|b(z)\|^2 + c(z)\right\} f(z) \quad (2.3.5)\end{aligned}$$

Proof: By (1.4) and definition of P we have:

$$\begin{aligned}\frac{\partial}{\partial s}\{Q_y(t,t-s)f\}(x) = Q_y(t,t-s)&\left\{\frac{L(q_y(t-s,.)f(.))}{q_y(t-s,.)} - \frac{Lq_y(t-s,.)}{q_y(t-s,.)}f(.)\right. \\ &\left. + \frac{LC_y(.)}{C_y(.)}f(.)\right\}(x),\end{aligned}$$

whence (2.3.1) follows from the fact that $\Delta(fg) = f\Delta g + g\Delta f + 2\langle\nabla f, \nabla g\rangle$, for real valued functions f and g.

Similarly :

$$\frac{\partial}{\partial t}\{Q_y(t,s)f\}(x) = \frac{L(q_y(t,.)(Q_y(t,s)f)(.))}{q_y(t,x)}(x) - \frac{Lq_y(t,x)}{q_y(t,x)}\{Q_y(t,s)f\}(x) + \frac{LC_y(x)}{C_y(x)}\{Q_y(t,s)f\}(x),$$

whence (2.3.2) follows as above.

Finally (2.3.3) follows from the same identity, (2.3.4) from the definitions of L and C_y, and (2.3.5) from (1.4.1).

(2.4) **Definition.**

Let L_O denote the smooth scalar second-order elliptic differential operator $L - c$. Then the q_y-transformed L_O-diffusion $x^t(u)$ starting from x is the time dependent diffusion with generator

$$L_O(z) + \nabla \log q_y(t-u, z) = \frac{1}{2}\Delta(z) + b(z) + \nabla \log q_y(t-u, z), \text{for } t > u \geq 0,$$

which may be represented as a strong solution of a stochastic differential equation.

As a consequence of the isometric embedding theorem of Nash, see Nash [1], we can obtain a smooth section X of the bundle $L(\underline{\mathbf{R}}^m; TM)$, for sufficiently large m, so that if $X^\star$ denotes the dual section of $L(T^\star M; \underline{\mathbf{R}}^m)$ then $A = XX^\star$ is the symbol of L. Then $x^t(u)$ is the solution of the following Itô differential equation with respect to the Levi-Civita connection:

$$dx^t(u) = X(x^t(u))\, dB(u) + \{b(x^t(u)) + \nabla \log q_y(t-u, x^t(u))\}\, du,$$
$$x^t(o) = x,$$

where $B(u)$ is m-dimensional Brownian motion .

(2.5) **Remark.** If b is a gradient vector field then from (2.3.5) we see that $x^t(u)$ is actually Brownian motion on M transformed with respect to r_y, i.e., it is the 'Brownian Riemannian Bridge Process' between x and y in time t, of Elworthy and Truman [1] and Elworthy [1] (Chapter IX §12D). So $x^t(u)$ tends to y almost surely as u increases to t and consequently the process does not explode.

For more general b we get a different process, but as we will see in (2.6), (1.4.2) means that the same argument as in Elworthy [1] (Chapter IX §12D) will show that it is a bridge process which is radially the same as the Euclidean Brownian bridge in $\mathbf{R}^n$: so in particular $x^t(u)$ tends to y almost surely as u increases to t and consequently this process does not explode either.

(2.6) Lemma. *The q_y-transformed L_0-diffusion $x^t(u)$ is a bridge process, i.e., $x^t(u)$ tends to y almost surely as u increases to t, whose radial component has the same law as the Euclidean Brownian bridge in $TM_y \cong \mathbf{R}^n$ between $\exp_y^{-1} x$ and 0 in time t . We will call this the 'Semi-Classical Bridge Process' between x and y in time t, associated with L as in Watling [1].*

Proof: Consider the function $R_y : M \to \mathbf{R}$ defined by $R_y(x) = d(x,y)$. Then we see that R_y is C^2 on $M \setminus \{y\}$ while for $n \geq 2$ and $x \neq y$ we have that almost surely $x^t(u)$ avoids y for $0 \leq u < t$. So we may apply Itô's formula for $0 \leq u < t$ to deduce that:

$$R_y(x^t(u)) = R_y(x) + \int_0^u \langle \nabla R_y(x^t(s)), X(x^t(s))dB(s)\rangle$$

$$+ \int_0^u \langle \nabla R_y(x^t(s)), b(x^t(s)) + \nabla \log q_y(t-s, x^t(s))\rangle \, ds + \int_0^u \frac{1}{2}\Delta R_y(x^t(s))\, ds$$

From (1.2.1) applied to R_y we see that:

$$\Delta R_y(z) = \frac{n-1}{R_y(z)} - 2\frac{\partial}{\partial r}\log \Theta_y(z).$$

From (1.4.1) and the observation that $\nabla E_y(z) = R_y(z)\nabla R_y(z)$ we see that:

$$\langle \nabla R_y(z), b(z) + \nabla \log B_y(z)\rangle = 0.$$

As $\|\nabla R_y(z)\|^2 = 1$ we see that:

$$\langle \nabla R_y(z), \nabla E_y(z)\rangle = R_y(z).$$

If we define:

$$W(u) = \int_0^u \langle \nabla R_y(x^t(s)), X(x^t(s))dB(s))\rangle,$$

then we see that it is a 1-dimensional Brownian motion from Elworthy [1] (Chapter V Corollary 5C), since if we define $H : M \to L(\underline{\mathbf{R}}^m, \underline{\mathbf{R}})$ by

$$H(z)(v) = \langle \nabla R_y(z), X(z)v\rangle$$

then

$$H^\star(z) = X^\star(z)(dR_y(z))$$

so we have

$$HH^\star = \|\nabla R_y(z)\|^2 = 1.$$

Thus denoting $R_y(x^t(s))$ by $r^t(s)$ we see that $r^t(s)$ satisfies:

$$r^t(u) = r^t(0) + W(u) + \frac{1}{2}(n-1)\int_0^u \frac{ds}{r^t(s)} - \int_0^u \frac{r^t(s)}{t-s}\, ds.$$

Consequently it is just the radial component of the Euclidean bridge in the statement of the Lemma. It then follows that $x^t(u)$ is a bridge process.

(2.7) **Proposition.** *Assume $\frac{LC_y}{C_y}$ is bounded above on M, then for f smooth of compact support and $t > s \geq 0$:*

$$\{Q_y(t,t-s)f\}(x) = E_x\left\{f(x^t(s))\exp\left(\int_0^s \frac{LC_y(x^t(u))}{C_y(x^t(u))}\,du\right)\right\}$$

where $x^t(u)$ is the semi-classical bridge process between x and y in time t.

(2.8) **Remark.**

This enables us to extend the domain of the $Q_y(t,s)$ to say bounded measurable functions f as the right-hand side of the above equality make sense for such functions. They then form a semigroup on this function space.

Proof: (Of (2.7)) Define $h : [t-s,t] \times M \times \mathbf{R} \to \mathbf{R}$ by

$$h(r,x,v) = v\{Q_y(r,t-s)f\}(x).$$

Notice this is smooth. Then:

$$\frac{\partial}{\partial r}h(r,x,v) = \{I_y(r)h(r,.,v)\}(x), \text{by (2.3.2)}.$$

Consider the process $y^t(r) = (\tau^t(r), x^t(r), v^t(r))$ on the domain of h given by:

$$\begin{aligned} d\tau^t(r) &= -dr \\ dx^t(r) &= X(x^t(r))\,dB(r) + \{b(x^t(r)) + \nabla \log q_y(\tau^t(r), x^t(r))\}\,dr \\ dv^t(r) &= \frac{LC_y(x^t(r))}{C_y(x^t(r))}v^t(r)\,dr \end{aligned}$$

with $x^t(0) = x$, $\tau^t(0) = t$, and $v^t(0) = 1$.

So:

$$\tau^t(r) = t - r,$$

and

$$v^t(r) = \exp\left\{\int_0^r \frac{LC_y(x^t(u))}{C_y(x^t(u))}\,du\right\}$$

So the result follows by applying Itô's formula to $h(y^t(s))$ observing the cancellation that occurs to get:

$$v^t(s)f(x^t(s)) = \{Q_y(t,t-s)f\}(x) + M(s),$$

where $M(s)$ is a martingale, with $M(0) = 0$. Then take expectations, observing that the martingale part must be bounded as the other non-constant term is by assumption, to deduce the result.

(2.9) **Theorem.** *Recalling the definitions in (1.1) we assume* $\frac{LC_y}{C_y}$ *is bounded above on* M*, then for* $t > 0$*:*

$$p(t,x,y) = q_y(t,x)E_x\left\{\exp\int_0^t \frac{LC_y(x^t(u))}{C_y(x^t(u))}\,du\right\},$$

where $x^t(u)$ *is the semi-classical bridge process between* x *and* y *in time* t*.*

Proof: Recalling (1.4), (2.1.1) and (2.6) simply let s tend to t in (2.7) and use dominated convergence for f a smooth function of compact support taking the constant value one in a neighbourhood of the geodesic segment between x and y.

(2.10) **Remark.**

This is just the elementary formula of Elworthy and Truman [1] and Elworthy [1] (Chapter IX §12 Theorem 12D) in the case that $b = 0$.

§3 Exact Formulæ Exhibiting Small-Time Asymptotic Behaviour

(3.1) **Definition.**

For $0 \le r \le s$ let $F(s,r)$ be the operator

$$\{F(s,r)f\}(z) = f(\gamma(s-r)),$$

where γ is the unique geodesic from z to y parameterised to take time s. These form a two parameter semigroup on, for example, bounded measurable functions.

(3.2) **Lemma.** *Assume* $\frac{LC_y}{C_y}$ *is bounded above on* M*, then for* f *smooth and of compact support:*

$$\frac{\partial}{\partial s}\{Q_y(t,t-s)F(t-s,t-r)f\}(x) = \{Q_y(t,t-s)(L_{C_y}(t-s)F(t-s,t-r)f)\}(x),$$

for $t > r \ge s \ge 0$*, where* L_{C_y} *is the operator defined by:*

$$L_{C_y}g(z) = \frac{L(C_y(.)g(.))(z)}{C_y(z)}.$$

Proof: Follows from (2.3.3) and definition of F(t-s,t-r).

(3.3) **Proposition.** *Assume $\frac{LC_y}{C_y}$ is bounded above on M, then for f a smooth function of compact support taking the constant value one in a neighbourhood of the geodesic segment between x and y, we have for any $N \geq 0$ and $0 \leq s < t$:*

$$\{Q_y(t, t-s)f\}(x) = 1 + a_1(s,x,y) + \cdots + a_N(s,x,y) + F_{N+1}(s,x,y),$$

where for $1 \leq n \leq N$:

$$a_n(s,x,y) = \int_0^s \cdots \int_0^{s_{n-1}} \Bigg\{ F(t, t-s_n) L_{C_y} F(t-s_n, t-s_{n-1}) L_{C_y} \cdots \\ F(t-s_2, t-s_1) \frac{LC_y(.)}{C_y(.)} \Bigg\}(x)\, ds_n \ldots ds_1,$$

and for $1 \leq n \leq N+1$:

$$F_n(s,x,y) = E_x \Bigg\{ \int_0^s \cdots \int_0^{s_{n-1}} \{L_{C_y} F(t-s_n, t-s_{n-1}) \cdots \\ L_{C_y} F(t-s_1, t-s) f\}(x^t(s_n)) \exp\left(\int_0^{s_n} \frac{LC_y(x^t(u))}{C_y(x^t(u))}\, du \right) ds_n \ldots ds_1 \Bigg\},$$

where $x^t(u)$ is the semi-classical bridge process between x and y in time t.

Proof: By (3.2) for any smooth h of compact support, and any $0 \leq r \leq s$:

$$\begin{aligned} \{Q_y(t,t-r)h\}(x) &= \{Q_y(t,t-r)F(t-r,t-r)h\}(x) \\ &= \{Q_y(t,t)F(t,t-r)h\}(x) \\ &\quad + \int_0^r \{Q_y(t,t-v)L_{C_y}F(t-v,t-r)h\}(x)dv \\ &= \{F(t,t-r)h\}(x) + \int_0^r E_x \Bigg\{ L_{C_y}[F(t-v,t-r)h](x^t(v)) \\ &\qquad \exp\left(\int_0^v \frac{LC_y(x^t(u))}{C_y(x^t(u))}\, du \right) \Bigg\}\, dv, \end{aligned}$$

by (2.6). So taking $h = f$, $r = s$ and $v = s_1$ we see the theorem is true for $N = 0$.

Then we proceed by induction taking:

$$h = L_{C_y} F(t-s_n, t-s_{n-1}) L_{C_y} \ldots L_{C_y} F(t-s_1, t-s) f,$$

which again is smooth of compact support, $r = s_n$ and $v = s_{n+1}$ from which we see that:

$$F_n(s,x,y) = a_n(s,x,y) + F_{n+1}(s,x,y)$$

and whence the result.

(3.4) **Definition.**

For $0 \leq r \leq s \leq 1$ let $G(s,r)$ be the operator defined as: $(G(s,r)f)(z) = f(\alpha(s-r))$, where α is the unique geodesic from z to y parameterised to take time $(1-r)$. These form a two parameter semigroup on, for example, bounded measurable functions.

(3.5) **Theorem.** *Recalling the definitions in (1.1) we assume* $\frac{LC_y(z)}{C_y(z)}$ *is bounded then we have for* $N \geq 0$, *provided that*

$$\left\{L_{C_y} G(r_{n-1}, r_n) \dots L_{C_y} G(r_1, r_2) \frac{LC_y(.)}{C_y(.)}\right\}(z),$$

for $0 \leq r_n \leq r_{n-1} \leq \dots \leq r_1 \leq 1$, *is bounded for* $2 \leq n \leq N+1$:

$$p(t,x,y) = q_y(t,x)\{1 + a_1(x,y)t + a_2(x,y)t^2 + \dots + a_N(x,y)t^N + R_{N+1}(t,x,y)t^{N+1}\}$$

where

$$a_1(x,y) = \int_0^1 \left\{G(r_1,0)\frac{LC_y(.)}{C_y(.)}\right\}(x)\, dr_1$$

and for $n \geq 2$:

$$a_n(x,y) =$$

$$\int_0^1 \dots \int_0^{r_{n-1}} \left\{(G(r_n,0) L_{C_y} G(r_{n-1},r_n) L_{C_y} \dots G(r_1,r_2)\frac{LC_y(.)}{C_y(.)}\right\}(x)\, dr_n \dots dr_1,$$

and where

$$R_1(t,x,y) = E_x\left\{\int_0^1 \frac{LC_y(x^t(tr_1))}{C_y(x^t(tr_1))} \exp\left(\int_0^{tr_1} \frac{LC_y(x^t(u))}{C_y(x^t(u))}\, du\right) dr_1\right\}$$

and for $n \geq 2$:

$$R_n(t,x,y) = E_x\left\{\int_0^1 \dots \int_0^{r_{n-1}} \left\{L_{C_y} G(r_{n-1},r_n) \dots G(r_1,r_2)\frac{LC_y(.)}{C_y(.)}\right\}(x^t(tr_n))\right.$$

$$\left.\exp\left(\int_0^{tr_n} \frac{LC_y(x^t(u))}{C_y(x^t(u))}\, du\right) dr_n \dots dr_1\right\}$$

where $x^t(u)$ *is the semi-classical bridge process between* x *and* y *in time* t.

Observe that $R_n(t,x,y)t^{n+1}$ *is* $o(t^n)$ *as* t *tends to* 0, *so the above formula gives the asymptotic expansion of* $p(t,x,y)$ *as* t *tends to* 0.

Proof:

Recalling (1.4) and (2.1.1) simply let s tend to t in (3.3) using dominated convergence for remainder term and finally change variables.

§4 Examples

(4.1) **Example.** *(Minakshishundaram-Pleijel Expansion)*

We calculate the first term from Theorem (3.5) of $p(t,y,y)$ in the case $c=0$:

$$a_1(y,y) = \int_0^1 \left\{ G(r,0)\frac{LC_y(.)}{C_y(.)} \right\}(y)\, dr_n \dots dr_1,$$
$$= \frac{LC_y(y)}{C_y(y)},$$

as $G(r,0) \equiv Identity$,

$$= \frac{1}{2}\Delta\Theta_y(y) - \frac{1}{2}\operatorname{div} b(y) - \frac{1}{2}\|b(y)\|^2,$$

as $(b + \nabla \log B_y)(y) = 0$ and $\Theta_y(y) = 1$,

$$= \frac{1}{12}S(y) - \frac{1}{2}\operatorname{div} b(y) - \frac{1}{2}\|b(y)\|^2,$$

where S is the scalar curvature, see Besse [1].

(4.2) **Example.** *(Hyperbolic n-space)*

For hyperbolic n-space with constant sectional curvatures $-R^{-2}$ we have:

$$\Theta_y(x) = \left(\frac{\frac{r}{R}}{\sinh(\frac{r}{R})}\right)^{\frac{n-1}{2}}$$
$$= \left(\frac{r}{\psi(r)}\right)^{\frac{n-1}{2}},$$

for $r = d(x,y)$, and $\psi(r) = \sinh(\frac{r}{R})$.
So by (1.2.1) applied to Θ_y we see that:

$$\frac{\Delta\Theta_y(x)}{2\Theta_y(x)} = \frac{-(n-1)\psi''(r)}{4\psi(r)} + \frac{(n-1)(n-3)}{8}\left\{\frac{1}{r^2} - \frac{\psi'(r)^2}{\psi(r)^2}\right\}$$
$$= \frac{-(n-1)}{4R^2} + \frac{(n-1)(n-3)}{8}\left\{\frac{1}{r^2} - \frac{1}{R^2\tanh^2(\frac{r}{R})}\right\}$$
$$= \frac{-(n-1)^2}{8R^2} + \frac{(n-1)(n-3)}{8}\left\{\frac{1}{r^2} - \frac{1}{R^2\sinh^2(\frac{r}{R})}\right\}$$

In particular when $n=3$ it is constant and so:

$$a_k(x,y) = (-2R^2)^{-k}\int_0^1 \cdots \int_0^1 dr_k \dots dr_1 = \frac{1}{k!}(-2R^2)^{-k},$$

which is just the kth term in the power series expansion of $\exp(\frac{-1}{2R^2})$, which is what you would expect from the exact formula in Theorem (2.7).

In general the scalar curvature S is given by, $\frac{-n(n-1)}{R^2}$. So observe that in this special case of hyperbolic n-space we have $\frac{\Delta\Theta_y(x)}{2\Theta_y(x)} = \frac{S(y)}{12}$, as pointed out in (4.1).

References.

Arede M.T. [1] (1983): *Géométrie du noyau de la chaleur sur les variétés. Thèse de Doctorat de 3ème Cycle Physique Théorique, Université D'aix Marseille, Faculté des Sciences de Luminy.*

Azencott R. et al [1] (1981): *Géodésiques et diffusions en temps petit. Séminaire de Probabilités, Université de Paris VII. Astérique 84–85 Société Mathématique de France.*

Azencott R. [2] (1984): *Densité des diffusions en temps petit, développements asymptotiques, Séminaire de Probabilité XVIII 1982/83. Lecture Note in Math. 1059, pp. 402–498: Springer Verlag.*

Besse A.L. [1] (1978): *Manifolds all of whose geodesics are closed. Ergebnisse der Mathematik 93: Springer Verlag.*

Bismut J.M. [1] (1984): *Large deviations and the Malliavin calculus. Progress in Mathematics 45, Birkhaüser Verlag.*

Elworthy K.D. [1] (1982): *Stochastic differential equations on manifolds. London Math. Soc. Lecture Notes series no. 70: Cambridge University Press.*

Elworthy K.D. and Truman A. [1] (1982): *The diffusion equation and classical mechanics: an elementary formula. In 'Stochastic Processes in Quantum Physics' pp. 136–146. ed. S. Albeverio et al, Lecture Notes in Physics no. 173: Springer-Verlag.*

Elworthy K.D., Ndumu M. and Truman A. [1] (1986): *An elementary inequality for the heat kernel on a Riemannian manifold and the classical limit of quantum partition function. In 'From Local Times to Global Geometry, Control and Physics' pp. 84–89, ed. K.D. Elworthy. Pitman Research Notes in Math. No. 150, Longman.*

Ikeda N. and Watanabe S. [1] (1986): *Malliavin calculus of Wiener functionals and applications. In 'From Local Times to Global Geometry, Control and Physics' pp. 132–178, ed. K. D. Elworthy. Pitman Research Notes in Math. No. 150, Longman.*

Kifer Y.I. [1] (1976): *On the asymptotics of the transition density of processes with small diffusion. Theory of Probability and its Applications XII no. 3.*

Molchanov S.A. [1] (1975): *Diffusion processes and Riemannian geometry. Russian Math. Surveys 30, 1–53.*

Nash [1] (1956): *The Imbedding Problem for Riemannian Manifolds. Ann. of Math. 63, 20–63.*

Ndumu M. [1] (1986): *An elementary formula for the Dirichlet heat kernel on Riemannian manifolds. In 'From Local Times to Global Geometry, Control and Physics' pp. 320–328, ed. K. D. Elworthy. Pitman Research Notes in Math. No. 150, Longman.*
Ndumu M. [2] (1987): *Ph.D. Thesis, University of Warwick, in preparation.*
Watling K.D. [1] (1986): *Formulæ for solutions to (possibly degenerate) diffusion equations exhibiting semi-classical and small-time asymptotics. Ph.D. Thesis, University of Warwick.*

STOCHASTIC MECHANICS FOR A POINT SOURCE

by

David Williams, Department of Pure Mathematics and Mathematical Statistics, University of Cambridge, 16 Mill Lane, Cambridge, CB2 1SB

Nick Steele and Aubrey Truman, Department of Mathematics and Computer Science, University College of Swansea, Singleton Park, Swansea, SA2 8PP

1. Introduction

The basic motivation comes from the following result.

LEMMA. Let ψ_E be a classical solution of $(-\frac{1}{2}\Delta + V)\psi_E = E\psi_E$, where $\psi_E = e^{R+iS}$ for real-valued R and S. Then $\tilde{\psi}_E^{\pm} = e^{R\pm S}$ is a classical solution of

$$(-\frac{1}{2}\Delta + \tilde{V})\tilde{\psi}_E = E\tilde{\psi}_E ,$$

where $\tilde{V} = V + |\nabla S|^2$, Bohm's effective potential.

The diffusion processes associated with the stationary-state solution ψ_E of the Schrödinger equation then have generators

$$(\tilde{\psi}_E^{\pm})^{-1}(-\tilde{H} + E)\tilde{\psi}_E^{\pm} , \quad \tilde{H} = -\frac{1}{2}\Delta + \tilde{V} .$$

The simplest example is provided by the case of point source in which $V \equiv 0$, $\psi_E = e^{ikr}/r$, $(-\frac{1}{2}\Delta + V)\psi_E = E\psi_E$, where $E = k^2/2$.

In this case,

$$\tilde{\psi}_E^{\pm}(x) = e^{\pm k|x|}/|x| ,$$

and the diffusion generators involved in picturing the motion of a particle emitted by the source are

$$(\tilde{\psi}_E^{\pm})^{-1}(\frac{1}{2}\Delta - \frac{1}{2}k^2)\tilde{\psi}_E^{\pm} = \frac{1}{2}\Delta + (\pm k - |x|^{-1})\frac{x}{|x|} \cdot \text{grad} .$$

Clearly, the relevant solution for a source should be that corresponding to $\tilde{\psi}_E^{+}$.

In accordance with the usual laws of quantum mechanics, the measure with density $|\psi_E|^2 = |x|^{-2}$ should be 'invariant for the system' in some appropriate sense.

In the language of probabilists, our aim then is to construct a system in which

(i) an individual particle emitted by the source performs a diffusion with infinitesimal generator

$$G_{\text{source}} = h^{-1}_{\text{source}} A h_{\text{source}} ,$$

where

$$h_{\text{source}}(x) = e^{k|x|}/|x| , \quad A = \tfrac{1}{2}\Delta - \tfrac{1}{2}k^2 ;$$

(ii) system has invariant density const. $|x|^{-2}$ on $\mathbb{R}^3\backslash\{0\}$.

The operator G_{source} is a Doob h-transform of the operator A . This note does two things:

a) it comments briefly on the Martin boundary theory associated with the operator A ,

b) it proposes two possible models (Models I and II) for the stochastic mechanics of the point source.

2. 'Spherically symmetric' theory. The radial part A^{rad} of A is given by

$$A^{\text{rad}} f = \tfrac{1}{2}f'' + r^{-1}f' - \tfrac{1}{2}k^2 f = \tfrac{1}{2}r^{-1}\left(\frac{d^2}{dr^2} - k^2\right)(rf) .$$

Of course, A^{rad} is the generator of '3-dimensional' Bessel process killed at rate $\frac{1}{2}k^2$. If h is harmonic for A^{rad}, then for some constants a and b ,

$$rh = ae^{kr} + be^{-kr} .$$

It is clear that extremal (or minimal) positive harmonic functions for A^{rad} are provided by h_0 and h_∞ , where

$$h_0(r) = e^{-kr}/r , \quad h_\infty(r) = (\sinh kr)/r .$$

In the language of Doob h-transforms,

$$G^{\text{rad}}_0 f = h_0^{-1} A^{\text{rad}}(h_0 f) = \tfrac{1}{2}f'' - kf'$$

gives the generator G^{rad}_0 of the A^{rad} diffusion conditioned to converge to 0 , and

$$G^{\text{rad}}_\infty f = h_\infty^{-1} A^{\text{rad}}(h_\infty f) = \tfrac{1}{2}f'' + (k\coth kr)f'$$

gives the generator of the A^{rad} diffusion conditioned to converge to ∞ .

The radial motion suggested by the stochastic mechanics has generator

$$G^{\text{rad}}_{\text{source}} f = h^{-1}_{\text{source}} A^{\text{rad}}(h_{\text{source}} f) = \tfrac{1}{2}f'' + kf' ,$$

and ignoring what happens at the boundary point 0 , this is the generator of Brownian motion with constant drift.

3. Brownian motion on $\mathbb{R}$ with constant drift. Let $\{u(t) : t \ge 0\}$ be a Brownian

motion on the *whole* of $\mathbb{R}$ with constant drift k and started at 0. Let $\sigma = \sup\{s : u(s) = 0\}$.

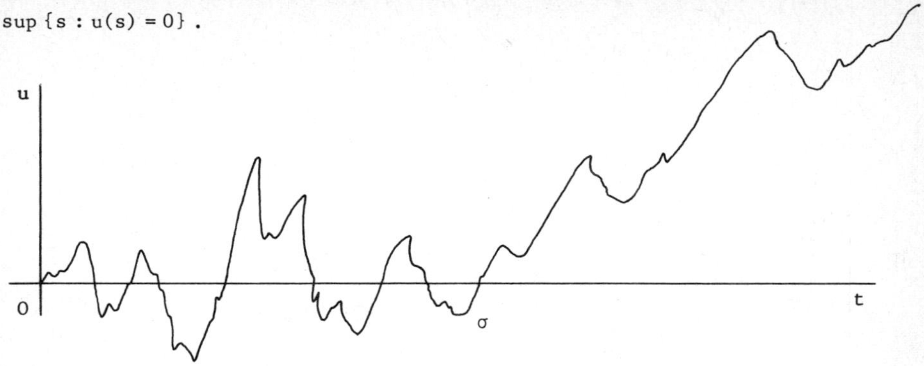

Then (Williams [2]) *the post-σ process* $\{u(t+\sigma) : t \geq 0\}$ *is a diffusion process with generator* G_∞^{rad}, and the pre-σ process $\{u(t) : t < \sigma\}$ is an independent diffusion with generator

$$\frac{1}{2}f'' - k\,\mathrm{sgn}(u)f' \quad \text{with boundary condition:} \quad f'(0+) - f'(0-) = 4kf(0)\,.$$

There are some nice 'superpositions of states' corresponding to h_0, h_∞ going on here!

Now of course we cannot have negative distances for the $\mathbb{R}^3$ model (unless we regard them as in an 'internal space within the source'). So we must prevent excursions below 0. There are two obvious ways to do this.

The first is simply to use a rubber to erase the excursions below 0 in the above picture, so that the particle stays at radial distance 0 for the times previously occupied by negative excursions. In this way, we get a *Type I model*:

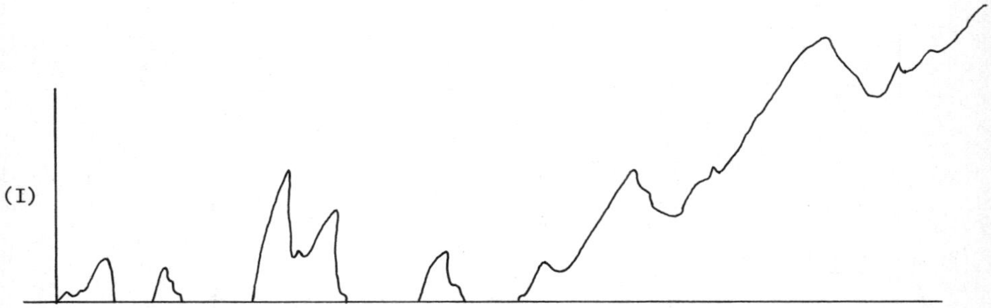

which is easily described rigorously in the language of *excursion theory*.

The second way (which physicists might prefer but which probabilists might not) is to close up the intervals in the (I) model to produce the *Type II picture*

(II)

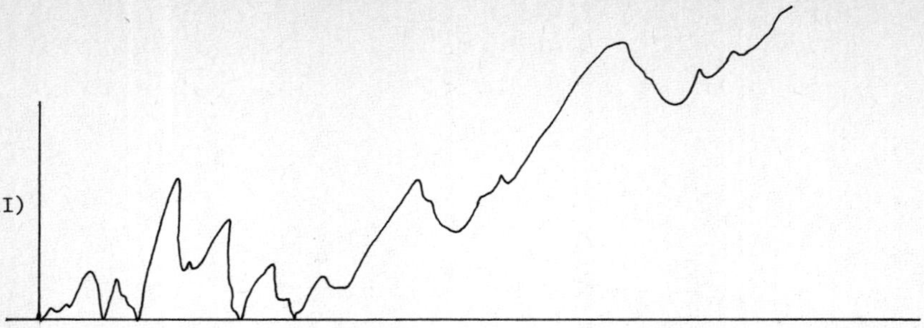

which (Cheers!) is Markovian, and is of course a diffusion with generator

$G^{rad}_{source} = \frac{1}{2} f'' + kf'$ with reflecting boundary condition: $f'(0) = 0$.

In this case II for the process starting at the origin 0 we can find the distribution of last exit times for B_a the open ball of radius a with centre at 0 and of first hitting times for the complement of B_a.

THEOREM

For the model II diffusion with $k > 0$ the transition density for the radial process is for $t,x,y > 0$

$$p_t(x,y) = e^{k(y-x) - \frac{k^2 t}{2}} (2\pi t)^{-\frac{1}{2}} \left(e^{\frac{-(x+y)^2}{2t}} + e^{\frac{-(x-y)^2}{2t}} \right) - k e^{2ky} \operatorname{erfc} \left\{ \frac{1}{\sqrt{2}} \left(\frac{x+y}{\sqrt{t}} + k\sqrt{t} \right) \right\}.$$

Further, setting $X(0) = 0$ and defining $\tau(a) = \inf\{s > 0 : X(s) \notin B_a\}$ and $\sigma(a) = \sup\{s > 0 : X(s) \in \bar{B}_a\}$, for each $t > 0$,

$$\mathbb{P}(\tau(a) < t) = 1 - \sum_n \frac{2\mu_n e^{-\frac{t}{2}\left(k^2 + \frac{\mu_n^2}{a^2}\right)} \sin \mu_n}{(a^2k^2 + \mu_n^2)((ak)^{-1}\cos^2 \mu_n + 1)},$$

where the sum is over $\mu_n > 0$, the successive roots of the equation in x : $ak \tan x = -x$, and

$$\mathbb{P}(\sigma(a) < t) = \int_0^t p_s(0,a)\, ds \Big/ \int_0^\infty p_s(0,a)\, ds = ke^{-2ka} \int_0^t p_s(0,a)\, ds .$$

It is difficult to see how the Schrödinger theory could yield correspondingly precise results for idealized point sources. A discussion of this and further results is given in Refs [1] and [4].

4. The Models (I) and (II) for a point source in $\mathbb{R}^3$. The source is postulated to contain an infinite number of particles. Each particle remains inside the source until a certain time called its 'trigger time'. The trigger times for the various particles form the events in a Poisson process on time-interval $(-\infty,\infty)$ of constant rate $c_1 = 4\pi ck$. Once triggered a particle will (independently of other particles) perform the Type I or Type II motion (all particles of course following the same Type). For any particle, its 'angular' motion on the unit shere S^2 is specified by the usual skew-product representation. Thus at any fixed number of time units after its trigger time, the angular part of the particle's position has the uniform distribution U on S^2, and, while away from 0 the angular part is $BM(S^2)$ run at rate $|x|^{-2}$.

THEOREM. For each of the two models, for every fixed time in $(-\infty,\infty)$, the distribution of particles in $\mathbb{R}^3\backslash\{0\}$ is a Poisson mass distribution with expectation density $c|x|^{-2}$.

This gives the appropriate interpretation of the fact that $|\psi_E|^2$ is an 'invariant density' for the system, an interpretation in full accord with Doeblin's interpretation for invariant measures of infinite total mass.

Since, for each particle,

$$\int_{\sigma+1}^{\infty} |x(t)|^{-2}\,dt < \infty \qquad (\text{because } |x(t)|/t \to 1),$$

the spherical clock will reach a limit, and so

$$\lim \frac{x(t)}{|x(t)|} \text{ exists in } S^2 .$$

5. Martin Boundary Theory

We conclude by explaining briefly how the distribution of limiting angular part can sometimes be calculated immediately by Martin boundary theory. We work with the generator

$$G_\infty = h_\infty(x)^{-1}(\tfrac{1}{2}\Delta - \tfrac{1}{2}k^2)h_\infty(x), \quad h_\infty(x) = (\sinh k|x|)/|x| .$$

The Green's function for this operator is

$$G(x,y) = h_\infty(x)^{-1} \frac{e^{-k|y-x|}}{2\pi|y-x|} h_\infty(y) .$$

For each y, we consider the Martin kernel function

$$K(x,y) = \frac{G(x,y)}{G(0,y)} .$$

Elementary trigonometry shows that if $\xi \in S^2$, and y_n is a sequence of points in $\mathbb{R}^3$ such that

$$|y_n| \to \infty, \quad y_n/|y_n| \to \xi ,$$

then

$$K(x,y_n) \to K(x,\xi) = \frac{k|x|}{\sinh k|x|} \exp(k\xi.x) .$$

We can immediately conclude that the Martin boundary for G_∞ is the 'sphere at ∞' which we represent in the obvious way by S^2. The functions $x \mapsto K(x,\xi)$ where $\xi \in S^2$ are the extremal positive harmonic functions for G_∞.

THEOREM (DOOB, HUNT). For each starting point x in $\mathbb{R}^3$ for a diffusion with generator G_∞, the limiting direction

$$L = \lim \frac{X(t)}{|X(t)|} \text{ exists in } S^2 .$$

Moreover,

$$\mathbb{P}[L \in d\xi | X(0) = x] = K(x,\xi)\mathbb{P}[L \in d\xi | X(0) = 0] .$$

But, since

$$\mathbb{P}[L \in d\xi | X(0) = 0] = U(d\xi) \qquad \text{(obviously!)} ,$$

we have

$$\mathbb{P}[L \in d\xi | X(0) = x] = \frac{k|x|}{\sinh k|x|} \exp(k\xi.x)U(d\xi) ,$$

the von-Mises distribution with pole at $x/|x|$ and concentration parameter $k|x|$.

See Williams [3] for an attempt to make Martin boundary theory simple.

References

[1] Steele, N., Ph.D. thesis, in preparation.

[2] Williams, D., Path decomposition and continuity of local time for 1-dimensional diffusions, I, Proc. London Math. Soc. Ser 3, 28, 738-68 (1974).

[3] —— Diffusions, Markov processes, and Martingales, Wiley, 1979.

[4] Williams, D., Steele, N., and Truman A., Stochastic Mechanics of Point Sources, in preparation.

COMPUTER STOCHASTIC MECHANICS

Kunio Yasue
Notre Dame Seishin University
Okayama 700, Japan*

and

Research Center Bielefeld-Bochum-Stochastics
University of Bielefeld
D-4800 Bielefeld 1, F. R. G.

*Permanent address

1. Introduction

A probabilistic formulation of quantum mechanics on path space has been introduced by Fényes and Nelson.[1-3] It has been called stochastic mechanics and attracted physicists' and mathematicians' attention gradually. As a physicist interested in stochastic mechanics for a relatively long time, it is a great pleasure for me to have an opportunity to address the present conference in Swansea. I wish to thank the organizing committee for it.

Because many mathematicians are present among the lecturers and in the audience, I will talk about the physicist' motivation for investigating stochastic mechanics. In quantum mechanics, as one might know, we have four different mathematical formulations; Schrödinger's wave mechanics, Heisenberg's matrix mechanics, Feynman's path integrals, and Fényes and Nelson's stochastic mechanics. They have been shown to be in agreement with each other, as far as effects in physical observables are concerned. However, they seem to show their own merits and demerits when applied to different physical systems. This is the very reason why stochastic mechanics should be investigated extensively. It may contain more physical information in certain situations than the other formulations. Indeed, as it offers the notion of path, it has been a conjecture that the sample paths of a stochastic process appearing in stochastic mechanics provide us with the trajectories of a quantum particle.[4]

One of the main dogmata of quantum mechanics is the impossibility to see the particle trajectories without disturbing the dynamical aspect. Perhaps, it may be true that stochastic mechanics is one of the few theoretical devices to see the unseen particle trajectories in quantum mechanics. If you could see the unseen object, you had gained an insight into the reality. Recent rapid development of electronic instruments with microcomputer in Japan makes it possible to fabricate a virtual apparatus with stochastic mechanics to see the unseen trajectories of quantum particle. I will show you some of the micro micro microphotographs taken by the virtual apparatus. Then, you may have an insight into the reality in quantum mechanics.[5]

Physicists seem to be radical enough, and for better understanding physical phenomena they do not hesitate to change their viewpoints drastically. Stochastic mechanics has suffered from unexpected distortions these years,[6] and a completely original probabilistic idea to understand quantum mechanics may again be needed. If such a new probabilistic framework of dynamics can be found, the original program of stochastic mechanics will be accomplished there and it will provide us with a deeper understanding of both probabilistic and dynamical aspects of quantum mechanics.[7,8]

2. Numerical Analysis in Stochastic Mechanics

Let us consider a quantum particle with mass equal to unity moving under the influence of a time-independent potential energy V. We put the Planck constant equal to 2π so that $\hbar = 1$. In Schrödinger's wavemechanics, we could think of the dynamics only in terms of the wavefunction ψ subject to the Schrödinger equation

$$i\frac{\partial\psi}{\partial t} = (-\frac{1}{2}\Delta + V)\psi. \tag{1}$$

Born's probabilistic interpretation claims that the absolute square of the wavefunction $|\psi|^2$ yields the probability density for the position of the particle. In stochastic mechanics, we make a hypothesis that the particle performs a stochastic process X subject to a forward stochastic differential equation

$$dX(t) = b(X(t),t)dt + dW(t), \tag{2}$$

where W is a forward Wiener process. The drift vector b is a dynamical variable suffering from a dynamical constraint through Newton's equation of motion in the mean

$$\frac{1}{2}(DD_*X(t) + D_*DX(t)) = -\nabla V(X(t)), \tag{3}$$

where D and D_* are respectively the mean forward and backward derivatives.[2,3] The probabilistic meaning of this equation is still not clear. It is used only as a partial differential equation satisfied by the drift vector b and the probability density ρ of the stochastic process X. In any case, the drift vector b in question is directly connected with the wavefunction ψ by the relation

$$b(x,t) = (\mathcal{R}e + \mathcal{I}m)\frac{\nabla\psi(x,t)}{\psi(x,t)}, \qquad (\psi(x,t) \neq 0). \tag{4}$$

The probability density ρ coincides with the absolute square of the wavefunction $|\psi|^2$, if the initial density does. These are essentially all that stochastic mechanics offers as a new mathematical framework of quantum mechanics. It seems relatively compact, and its impact upon probability theory is not so strong. However, its impact upon quantum theory is so radical that even today many physicists take no notice of it. The reason is that stochastic mechanics incorporates the notion of particle trajectory into quantum mechanics and seems to provide us with a naive picture of particle dynamics in atomic scale. It is in clear contrast with the so-called Copenhagen interpretation and further careful investigations from the physical point of view will be needed before it may be accepted by a majority of physicists. Despite those fundamental questions remaining still open, it will be quite interesting from an intuitive point of view to test whether the sample paths of the stochastic process X well represent the quantum dynamics of the particle or not. If it hits one's better intuition, then the conjecture that the sample paths of X provide us with the unseen trajectories of a quantum particle should be considered seriously. There will be two different approaches to see the sample paths of stochastic processes X's appearing in stochastic mechanics, that is, analytic and numerical ones. I leave the analytic approach to mathematicians, and will talk about the numerical one. It is true that we cannot see infinitely many sample paths by computer, and therefore our numerical approach tells us only qualitative aspects of the stochastic process. However, certain dynamical aspect of the quantum trajectory of a particle may be extracted from it.

In relatively simpler cases of quantum dynamics, we can find explicit forms of wavefunction ψ's or at least some good approximations. If this is the case, we start from the forward stochastic differential equation (2) with the drift vector b given by Eq. (4) in terms of the known form of wavefunction ψ. In order to put the forward stochastic differential equation (2) into computer, a forward difference equation is introduced. Namely, we wish to compute the time series of sample values of position for integer multiples of the mesh of time $\Delta t > 0$,

$$X(0), \quad X(\Delta t), \quad X(2\Delta t), \quad \ldots, \quad X(j\Delta t), \quad \ldots . \tag{5}$$

This time series can be computed by the iteration with the forward difference equation

$$X((j+1)\Delta t) = X(j\Delta t) + b(X(j\Delta t), j\Delta t)\Delta t + W((j+1)\Delta t) - W(j\Delta t), \tag{6}$$

for $j = 0,1,2,\ldots$. The increment of the forward Wiener process in the right-hand side is a Gaussian random variable with mean 0 and covariance $\sqrt{\Delta t}$. Samples of such a Gaussian random variable can be generated as a sum of many uniformly distributed pseudo random numbers. In this way, we can assign by computer a definite sample value to the right-hand side of Eq. (6) in each step of iterations. The time series (5) thus obtained approximates a sample path of the stochastic process X by linear interpolation.[5]

It is difficult to understand the dynamical property of a stochastic process by regarding only one sample path. For the better understanding, we need to see as many sample paths as possible. Then, the totality of those computer generated sample paths of the stochastic process X should be taken into account as a qualitative representative of the unseen quantum trajectory of a particle. Now we will proceed to the case study of quantum dynamics.

3. Case Study 1 (Superposition)

One of the most strange concepts in quantum mechanics is the superposition. Each wavefunction subject to the Schrödinger equation (1) carries each dynamics of a quantum mechanical particle. Given two wavefunctions ψ_1 and ψ_2, say, we can always get another one by a linear combination of them.

$$\psi = a_1\psi_1 + a_2\psi_2,$$

where a_1 and a_2 are complex numbers chosen such that $|\psi|^2 = 1$. We call ψ a superposition of ψ_1 and ψ_2. Since the new wavefunction ψ also satisfies the Schrödinger equation (1), it certainly carries quantum dynamics of the particle. However, dynamical meaning of the superposition ψ is obscure. In the orthodox interpretation of quantum mechanics, it is used to be understood as follows: suppose ψ_1 and ψ_2 are mutually orthogonal proper wavefunctions of a self-adjoint operator P representing a certain dynamical variable. Let p_1 and p_2 be their proper values. Then, we will find the value p_1 with probability $|a_1|^2$ and the value p_2 with probability $|a_2|^2$, if the dynamical variable P is measured in the superposition ψ. As long as the dynamical variable P is concerned with, the superposition ψ looks like ψ_1 with probability $|a_1|^2$ and ψ_2 with $|a_2|^2$. This is not the case, however, for other dynamical variables incompatible with P. In the superposition ψ, there exists a kind of intereference between ψ_1 and ψ_2. This can be seen, for example, by computing the probability density,

$$\begin{aligned} \rho &= |\psi|^2 \\ &= |a_1\psi_1 + a_2\psi_2|^2 \\ &= |a_1|^2|\psi_1|^2 + |a_2|^2|\psi_2|^2 + a_1\bar{a}_2\,\psi_1\bar{\psi}_2 + \bar{a}_1 a_2\,\bar{\psi}_1\psi_2. \end{aligned} \tag{8}$$

The last two terms in the third right-hand side of Eq. (8) manifest the interference of dynamcics. It seems difficult to extract some dynamical meaning of superposition from the complex expression (8). Therefore, it will be of interest to compare the sample paths of the stochastic process X associated with the superposition ψ with those of stochastic processes X_1 and X_2 associated respectively with ψ_1 and ψ_2. Dynamical aspect of the superposition ψ, then, will be less obscure. We assume a_1 and a_2 to be real, for simplicity. Let us write down the polar decompositions of ψ_1, ψ_2, and ψ,

$$a_1\psi_1 = exp(R_1 + i\,S_1), \tag{9}$$

$$a_2\psi_2 = exp(R_2 + i\,S_2), \tag{10}$$

$$\psi = exp(R + i\,S). \tag{11}$$

Then, it is immediate to see R and S in terms of R_1, R_2, S_1 and S_2, obtaining

$$R = \frac{1}{2}\,log\left[2e^{R_1+R_2}[cosh(R_1 - R_2) + cos(S_1 - S_2)]\right], \tag{12}$$

$$S = arctan\left(\frac{e^{R_1}sin\,S_1 + e^{R_1}sin\,S_2}{e^{R_1}cos\,S_1 + e^{R_2}cos\,S_2}\right). \tag{13}$$

The stochastic processes X_1, X_2, and X associated with the wavefunctions ψ_1, ψ_2, and ψ, respectively, satisfy the forward stochastic differential equations

$$dX_1(t) = b_1(X_1(t),t)\,dt + dW(t)\,, \tag{14}$$

$$dX_2(t) = b_2(X_2(t),t)\,dt + dW(t)\,, \tag{15}$$

$$dX(t) = b(X(t),t)\,\mathrm{dt} + dW(t)\,, \tag{16}$$

where the drift vectors b_1, b_2, and b are given respectively by

$$b_1 = \nabla(R_1 + S_1), \tag{17}$$

$$b_2 = \nabla(R_2 + S_2), \tag{18}$$

$$b = \nabla(R + S), \tag{19}$$

By Eqs. (12), (13), and (19), we may understand how complex the dynamics in the superposition ψ is. The drift vector b shows highly nonlinear and complicated dependences on the drift vectors b_1 and b_2.

In order to extract some dynamical and probabilistic aspects of superposition, we consider here the simplest case of a free particle in one dimension. Let $\psi_1 = exp\{+ipx-i(p^2/2)t\}$ be an outgoing wave, and $\psi_2 = exp\{-ipx-i(p^2/2)t\}$ be an incoming wave. We put $a_1 = \sqrt{r}$ and $a_2 = \sqrt{1-r^2}$ for a positive constant r, and introduce the superposition $\psi = a_1\psi + a_2\psi$. For the equal weight superposition with $r = \frac{1}{2}$, ψ becomes a standing wave proportional to $cos(px)$. In the orthodox interpretation of quantum mechanics, ψ_1 and ψ_2 represent each particle dynamics moving freely with constant momenta p and $-p$, and the superposition ψ with $r = \frac{1}{2}$ represents a particle dynamics standing in a local spatial region. Let us see the sample paths of X_1, X_2, and X. In figures 1, notice that the outgoing and incoming waves ψ_1 and ψ_2 carry uniform probability densities $\rho_1 \propto 1$ and $\rho_2 \propto 1$, whereas the standing wave ψ with $r = \frac{1}{2}$ carries a periodic probability density $\rho \propto cos^2(px)$.

The totality of those computer generated sample paths well manifests the qualitative dynamical aspects of particle trajectories associated respectively with outgoing wave ψ_1, incoming wave ψ_2, and standing wave ψ.

What about the dynamics in the general superposition ψ with $r \neq \frac{1}{2}$? Figures 2 show the sample paths of X_1, X_2, and X, for r = 0.1, 0.2, 0.3, 0.4, 0.5, 0.6,

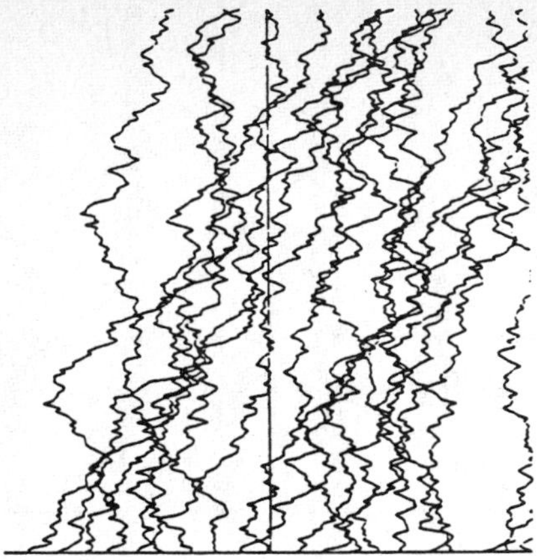

Fig.1a

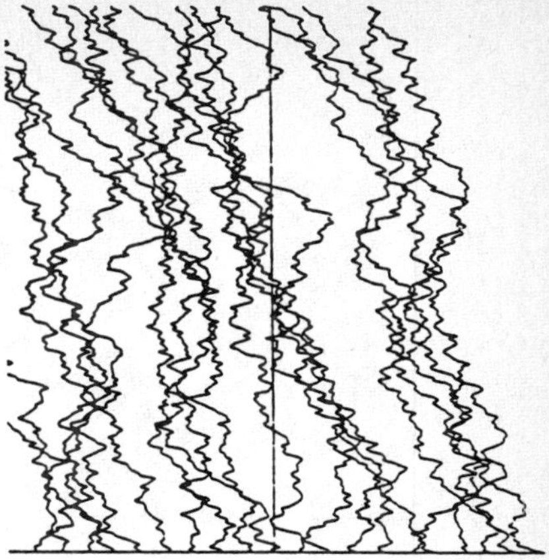

Fig.1b

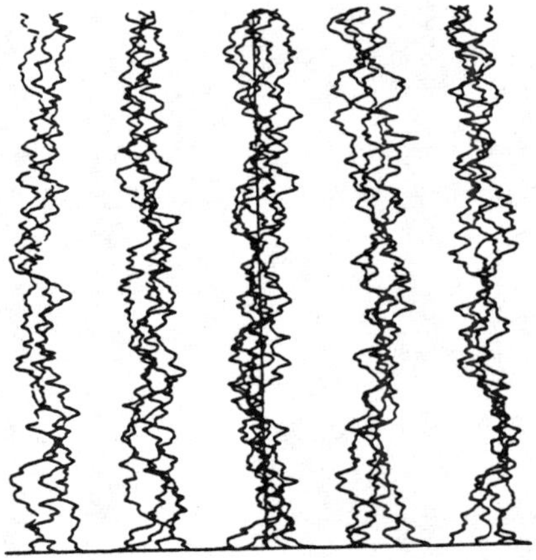

Fig.1c

Fig. 1a~1c. Sample paths of a quantum mechanical free particle associated with the outgoing wave ψ_1 (Fig. 1a), the incoming wave ψ_2 (Fig. 1b), and the standing wave $\psi = \psi_1 + \psi_2$ (Fig. 1c). Numerical factors are chosen as $p = 1.0$, t-range $= 5.0$, and x-range $= 7.5$.

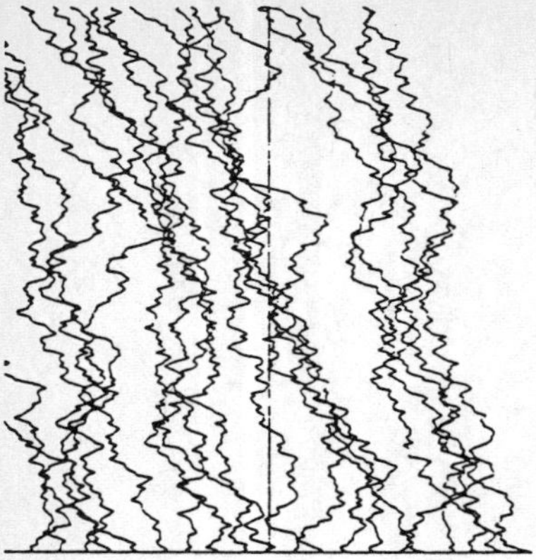
Fig.2a

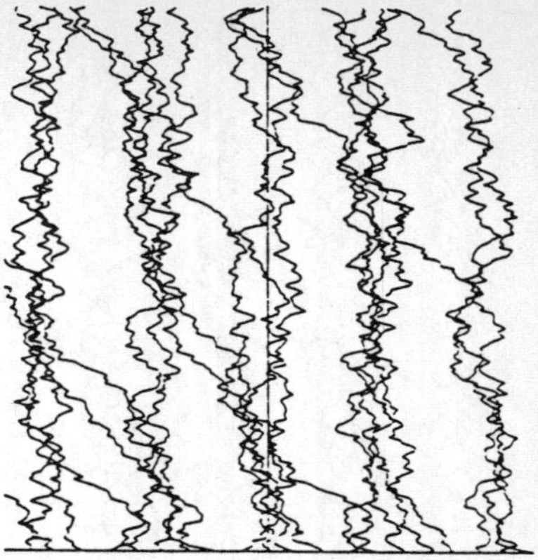
Fig.2b

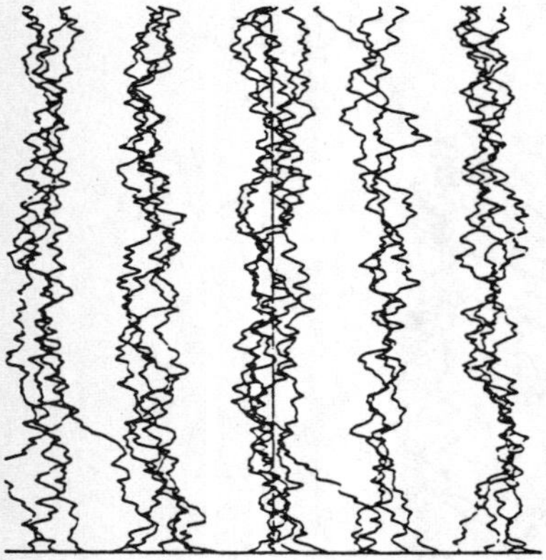
Fig.2c

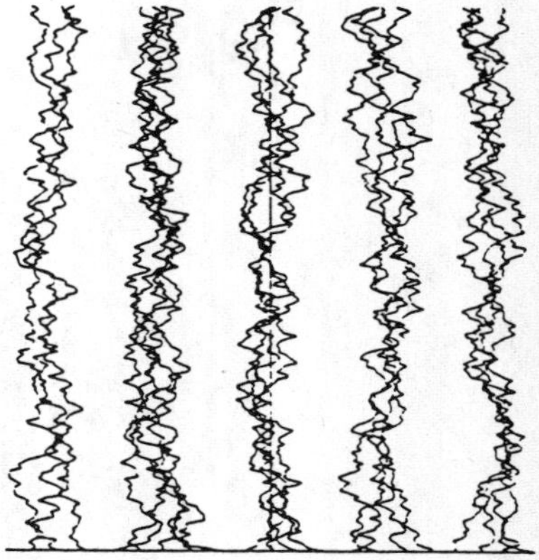
Fig.2d

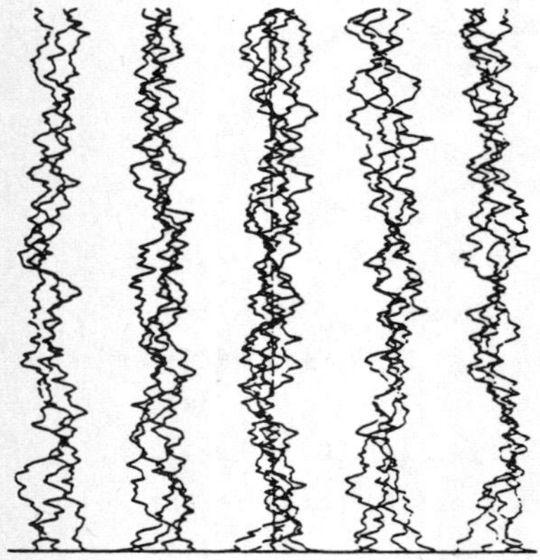
Fig.2e

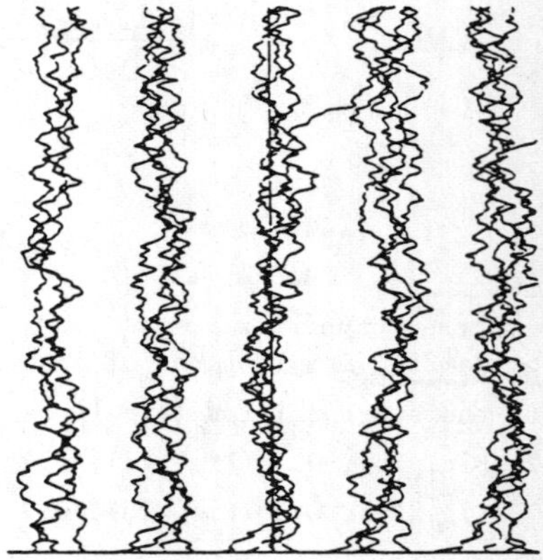
Fig.2f

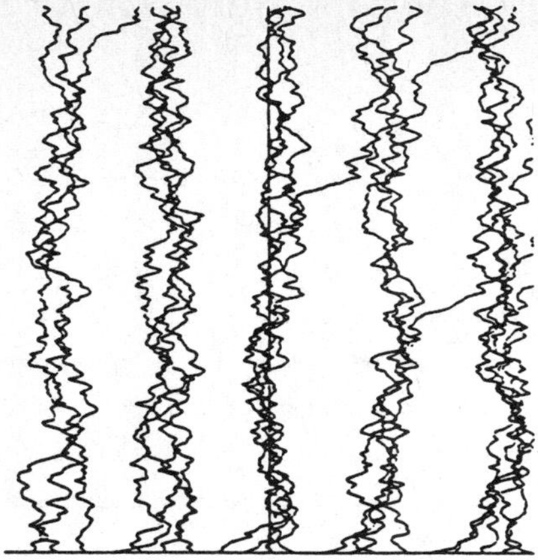

Fig.2g

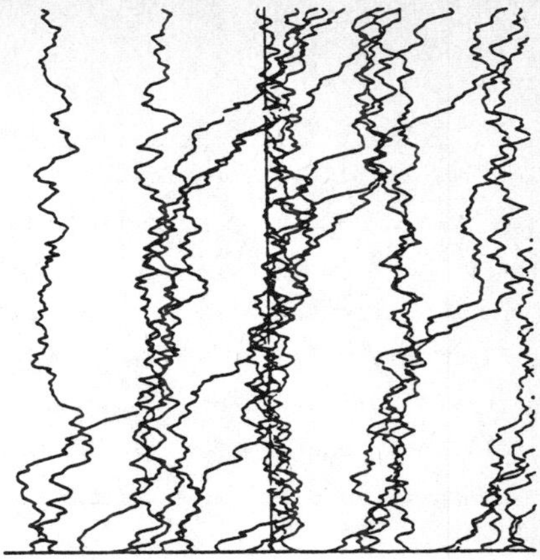

Fig.2h

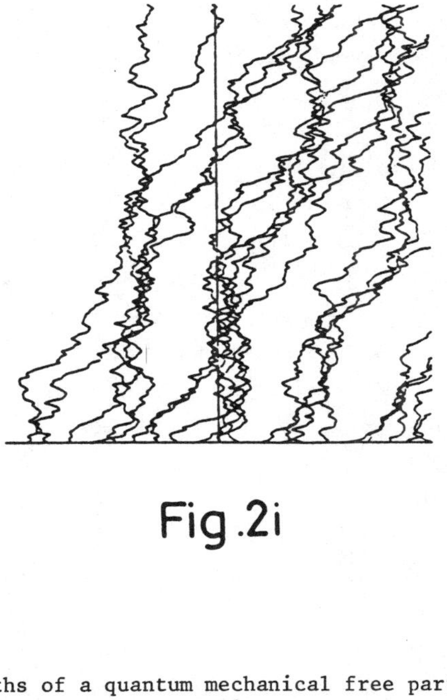

Fig.2i

Fig. 2a∿2i. Sample paths of a quantum mechanical free particle associated with the superposition $\psi = \sqrt{r}\,\psi_1 + \sqrt{(1-r^2)}\,\psi_2$, for r = 0.1 (Fig. 2a), 0.2 (Fig. 2b), 0.3 (Fig. 2c), 0.4 (Fig. 2d), 0.5 (Fig. 2e), 0.6 (Fig. 2f), 0.7 (Fig. 2g), 0.8 (Fig. 2h), 0.9 (Fig. 2i). Numerical factors are the same as Fig. 1.

0.7, 0.8, 0.9. There, you will see that the more you have ψ_1 in the superposition, the more you find sample paths going away from left to right.

4. Case Study 2 (Two Slit Interference)

Let us recall the physical setting of the famous two-slit interference thought experiment. A quantum particle is emitted from a certain source and it reaches a certain point of a detecting film later on. Between the source and the detecting film we place an infinite plate with two parallel slits separated by a small distance. The probability distribution of the particle on the detecting film, that can be obtained by successive emissions of the particles, shows the interference pattern when the two slits are open and one does not observe through which one the particle goes. Once one observes that the particle goes through one of the slits, the interference pattern disappears.

In the orthodox interpretation of quantum mechanics, we cannot say anything about the particle trajectory except the probability distribution of position, though it is indeed a particle that hits the detecting film. This prohibited us from asking through which slit passed the particle. It will be quite interesting, therefore, to revisit the problem of two-slit interference from the point of view of stochastic mechanics. Since then we can illustrate the unseen trajectories of a particle. Nelson investigated it analytically in his recent book,[3] and we will use the same notation as he.

We wish to claim that the wavefunction does go through both of the two slits but not the particle itself, though it is impossible to know which is crossed. Indeed, we cannot observe a particle passing through a slit without giving fatal damage to the wavefunction. As the wavefunction guides the particle where it has to go, the dynamics of the particle after the measurement does differ from the original one so that the interference disappears. It seems worthwhile to see the conceptual difference between the physical conditioning by measurements and the mathematical conditioning by constraining sample paths. The former does touch the wavefunction, whereas the latter does not. As Zambrini and Nelson have shown,[3,7,8] the probability distribution of the particle with interference is precisely the sum of the conditioned probability distributions, one conditioned to pass through the upper slit and the other the lower one. The concept of probability and the Bayes rule in quantum mechanics do not differ from the conventional ones.

It has been believed by many physicists that the concept of probability is completely different from the conventional one only because the probability distribution with interference cannot be expressed as a sum of the probability distributions conditioned physically to pass through each of the two slits. Physical conditioning such as to close the lower slit modifies the wavefunction and so the underlying stochastic process. Bayes rule should be satisfied only when we assume the mathematical conditioning that does not modify the wavefunction. It is not proper to modify the good old concept of probability in order to avoid the real

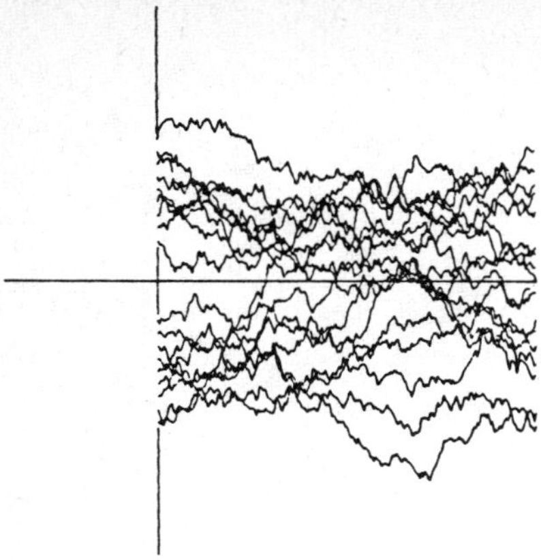

Fig.3a

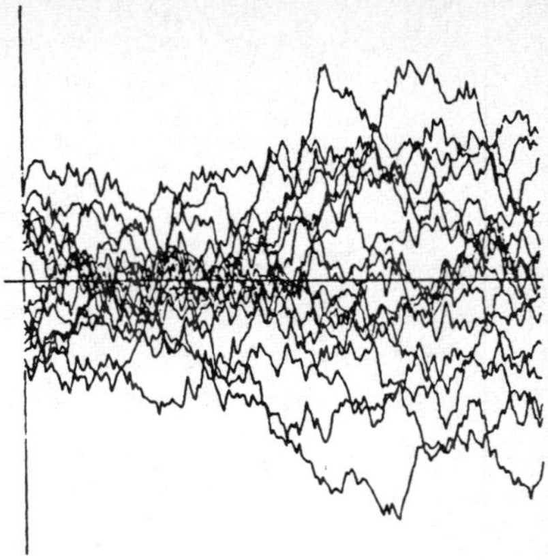

Fig.3b

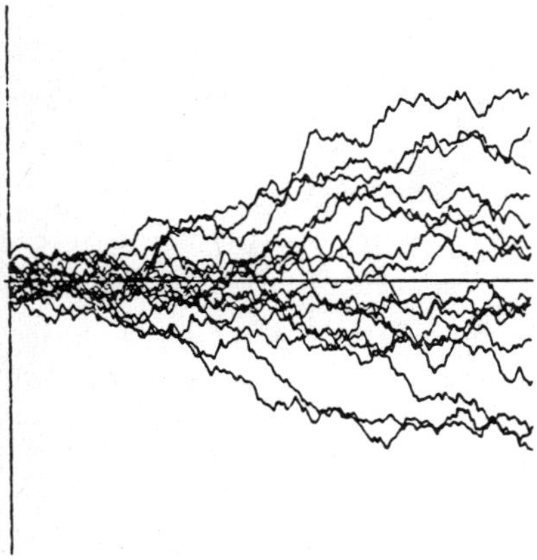

Fig.3c

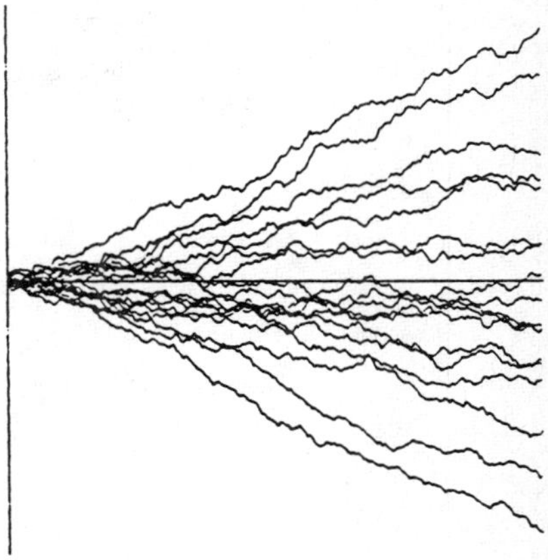

Fig.3d

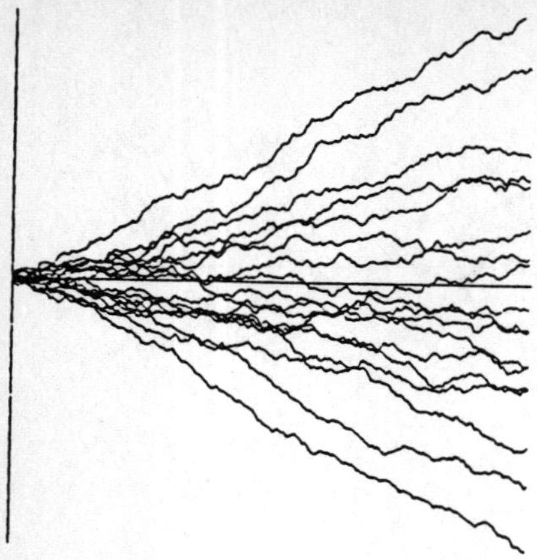

Fig.3e

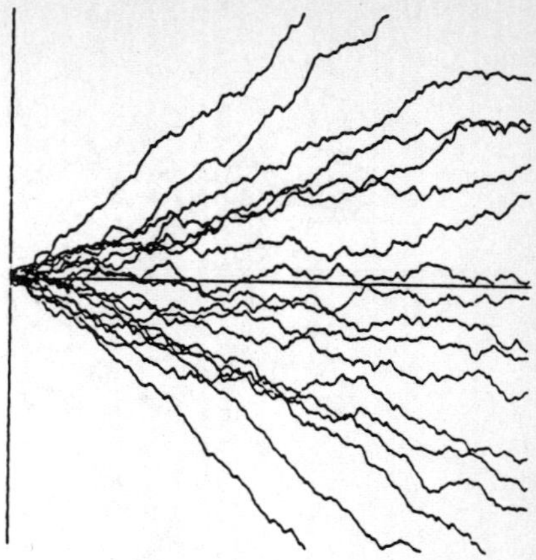

Fig.3f

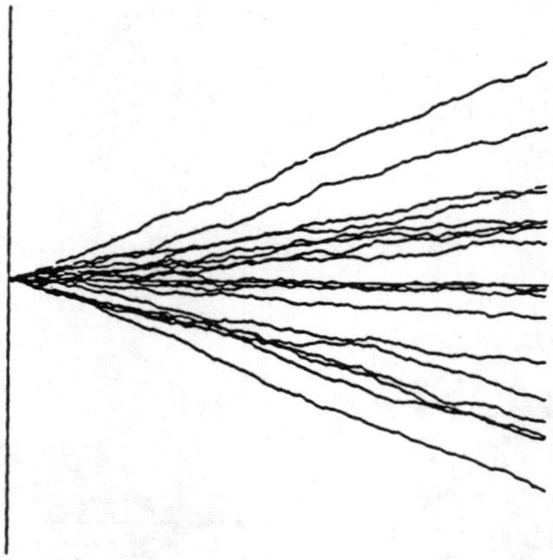

Fig.3g

Fig. 3a∿3g. Sample paths of a quantum mechanical free particle passing through the two slits. Separation of the two slits is 1.0, and the width of each slit is 1.25. Time scale is chosen to be 0.5 (Fig. 3a), 5.0 (Fig. 3b), 20.0 (Fig. 3c), 90.0 (Fig. 3d), 120.0 (Fig. 3e), 200.0 (Fig. 3f), 500.0 (Fig. 3g).

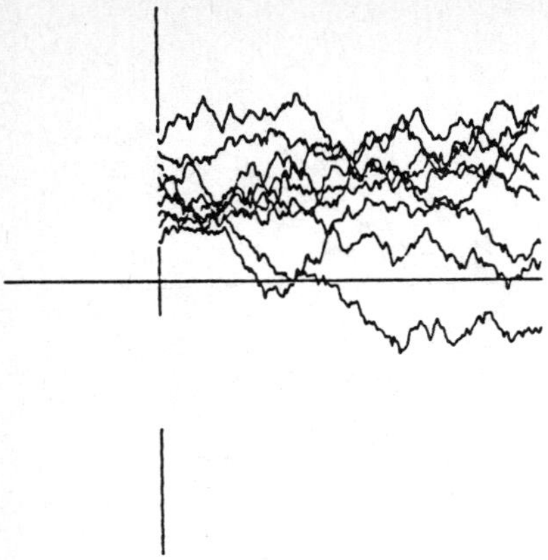

Fig.4a

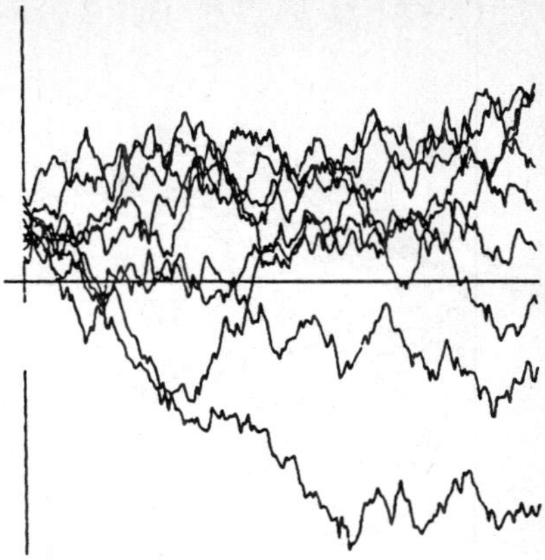

Fig.4b

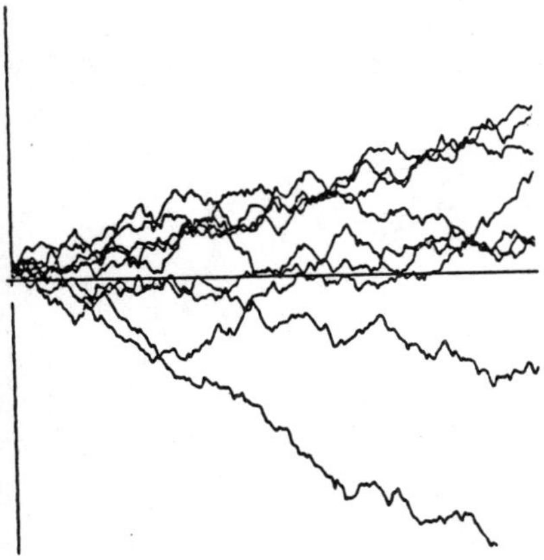

Fig.4c

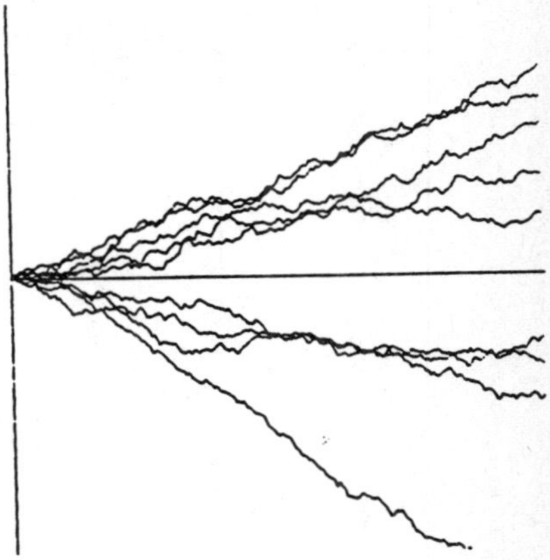

Fig.4d

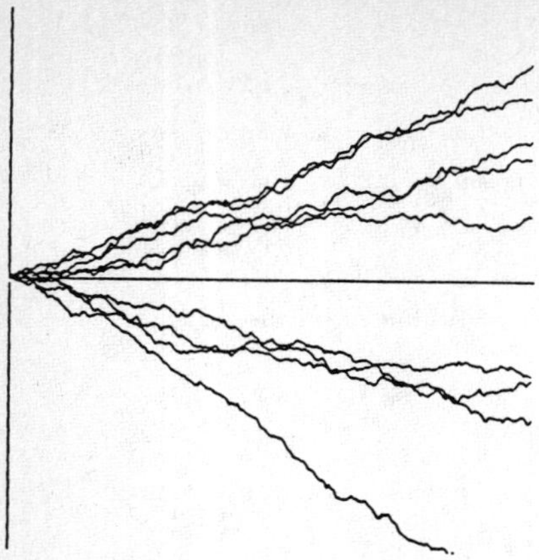

Fig.4e

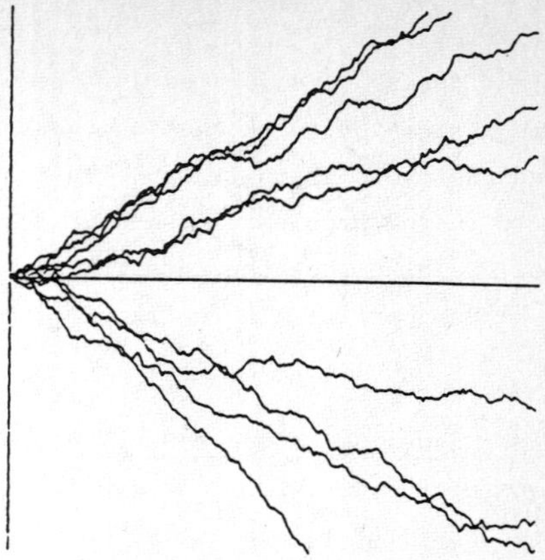

Fig.4f

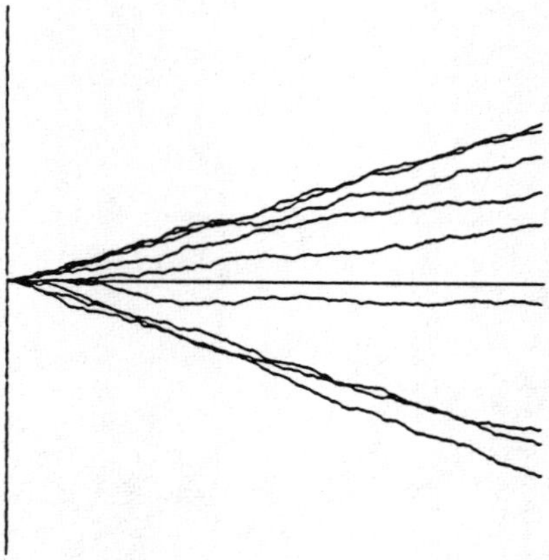

Fig.4g

Fig. 4a∿4g. Sample paths of a quantum mechanical free particle conditioned to pass through the upper slit. Numerical factors are the same as Fig. 3.

difficulties.

In figures 3 and figures 4 we see the totality of sample paths of a stochastic process associated with the superposition of two Gaussian wavefunctions passing through the two slits. It may illustrate the unseen quantum trajectories of the particle in the problem of two-slit interference. We show it for various time scales. We also show the totality of sample paths conditions mathematically to pass through the upper slit. The complicated dynamics of quantum interference will be found there.

5. Case Study 3 (Harmonic Oscillators)

One of the most familiar dynamical systems in physics is a one-dimensional harmonic oscillator. Since the potential energy is of quadratic form

$$V = \frac{1}{2}x^2, \tag{20}$$

the Schrödinger equation (1) can be integrated. Especially, we have a series of proper wavefunctions ψ_n's such that $\psi_n(x,t) = u_n(x)\, exp\{-i(n+\frac{1}{2})t\}$ and u_n solves the proper value problem

$$\left(-\frac{1}{2}\frac{d^2}{dx^2} + \frac{1}{2}x^2\right)u_n(x) = \left(n+\frac{1}{2}\right)u_n(x), \tag{21}$$

for $n = 0,1,2,\ldots.$ We call ψ_0 the ground state, ψ_1 the first excited state, ψ_2 the second excited state, and so on. Those proper wavefunctions ψ_n's can be obtained explicitly in terms of Hermite polynomials with a common Gaussian weight. For example, we have

$$\psi_0(x,t) = \left(\frac{1}{\pi}\right)^{\frac{1}{4}} e^{-\frac{1}{2}x^2 - i\frac{1}{2}t}, \tag{22}$$

$$\psi_1(x,t) = \left(\frac{4}{\pi}\right)^{\frac{1}{4}} x e^{-\frac{1}{2}x^2 - i\frac{3}{2}t}, \tag{23}$$

$$\psi_2(x,t) = \left(\frac{1}{4\pi}\right)^{\frac{1}{4}} (x^2-1) e^{-\frac{1}{2}x^2 - i\frac{5}{2}t}. \tag{24}$$

They are also called stationary states, since the probability densities are time-independent.

From the point of view of stochastic mechanics, the dynamics of a quantum mechanical harmonic oscillator in the stationary state ψ_n is given by the stochastic process X with invariant probability density $|\psi_n|^2$. For the better understanding of the dynamical aspect of stationary states of the harmonic oscillator, figures 5 exhibit the totality of computer generated sample paths of the stochastic processes X's associated with the ground state ψ_0, the first excited state ψ_1, and the second excited state ψ_2, respectively. We will find there the segregation and mutual balancing of sample paths.

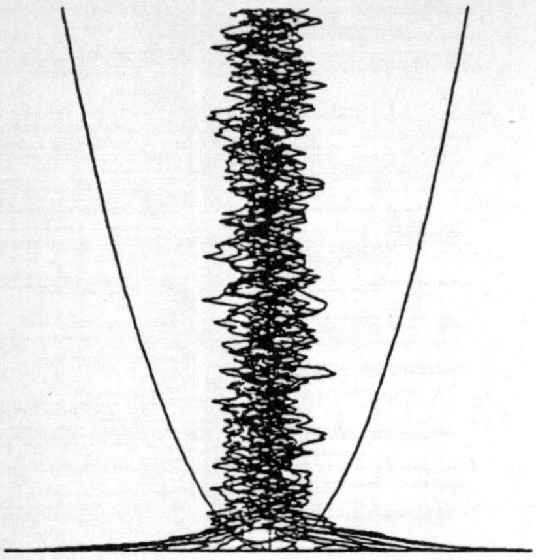

Fig.5a

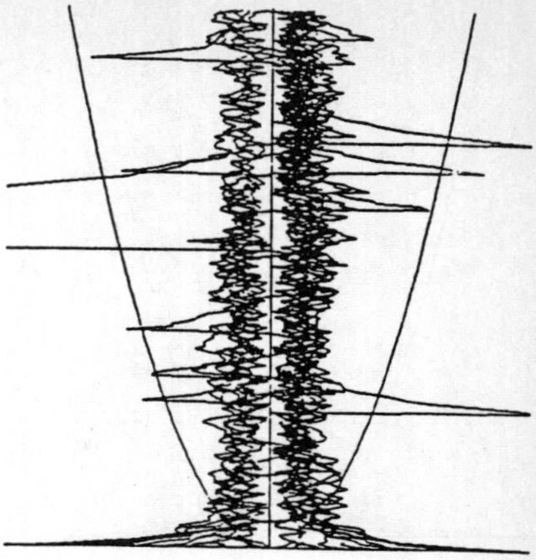

Fig.5b

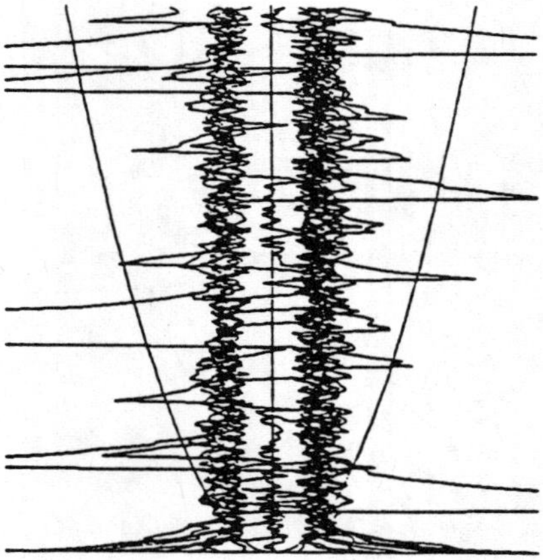

Fig.5c

Fig. 5a∿5c. Sample paths of a quantum mechanical harmonic oscillator in the ground state ψ_0 (Fig. 5a), in the first excited state ψ_1 (Fig. 5b), and in the second excited state ψ_2 (Fig. 5c). Time scale is chosen to be 10.0.

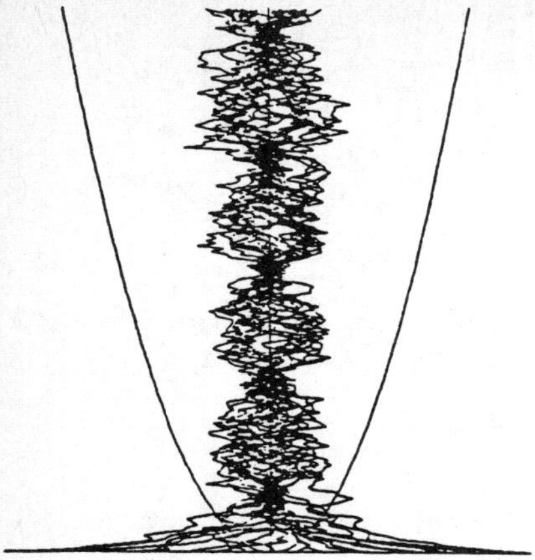

Fig.6a

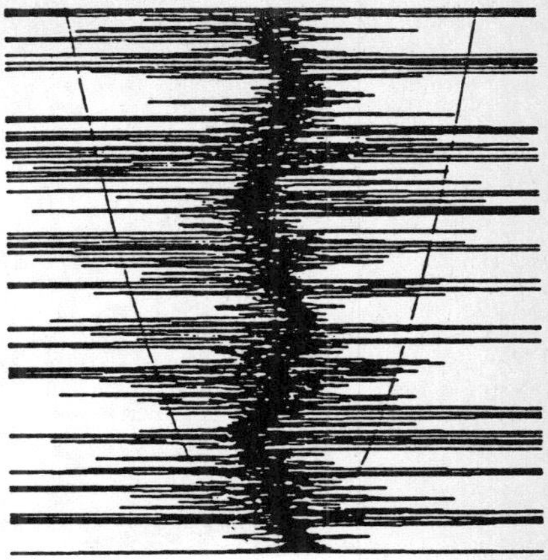

Fig.6b

Fig. 6a,6b. Sample paths of a quantum mechanical harmonic oscillator in the superpositions $\psi = \psi_0 + \psi_1$ (Fig. 6a) and $\psi = \psi_0 + \psi_1 + \psi_2$ (Fig. 6b).

It has been a physical conjecture that the more you superpose the higher excited states, the more the quantum trajectories approach the classical ones. To see whether this conjecture seems reasonable or not, we will compute the sample paths of the stochastic processes associated with the superpositions

$$\psi' = \psi_0 + \psi_1, \tag{25}$$

$$\psi'' = \psi_0 + \psi_1 + \psi_2, \tag{26}$$

respectively, and exhibit them in figures 6. The computer generated sample paths seem to claim the validity of this conjecture. Most of the sample paths gather into the vicinity of a classical oscillatory trajectory.

6. Conclusions

A numerical analysis with computer of the quantum mechanical trajectory of a particle has been proposed within the realm of stochastic mechanics. It has been shown that the present approach with computer graphics illustrates the typical samples of the unseen trajectory of a particle in quantum mechanics. Intuitive meanings of fundamental concepts in quantum mechanics such as the superposition and the interference have been clarified.

Acknowledgements

The author would like to take the opportunity to thank Professor Aubrey Truman for a very kind invitation to participate in the conference. He also thanks Professor Jean-Claude Zambrini for helpful advices and correspondences, Professor Tetsuya Misawa for intensive discussions, and Miss Mihoko Monobe for skillful assistance in preparing the manuscript. Computation was performed at the Research Institute for Informatics and Science (RIIS) of Notre Dame Seishin University with TOSBAC UX-700. The hospitality and support of the Research Center BiBoS of the University of Bielefeld are gratefully acknowledged.

References

1. I. Fényes, Z. Phys. 132, 81 (1952).
2. E. Nelson, Dynamical Theories of Brownian Motion, (Princeton University Press, Princeton, 1967).
3. E. Nelson, Quantum Fluctuations, (Princeton University Press, Princeton, 1985).
4. A. Truman and J.T. Lewis, "The Stochastic Mechanics of the Ground-State of the Hydrogen Atom", BiBoS preprint 37/84, to appear in Proceedings BiBoS I.
5. K. Yasue and J.-C. Zambrini, Ann. Phys. (N.Y.) 159, 99 (1985).
6. E. Nelson, "Field Theory and the Future of Stochastic Mechanics", BiBoS preprint 173/85, to appear in Proc. Ascona Conference "Stochastic Processes in Classical and Quantum Systems".
7. J.-C. Zambrini, Phys. Rev. A33, 1532 (1986).
8. J.-C. Zambrini, "Variational processes and stochastic versions of mechanics", to appear in J. Math. Phys.

NEW PROBABILISTIC APPROACH TO THE CLASSICAL HEAT EQUATION

J.C. Zambrini
Mathematics Institute
University of Warwick
Coventry CV4 7AL

1. Historical survey

To look at the history of the theories of Brownian motion is a fascinating exercise in which the indispensable interaction between theoretical physics and mathematics is illustrated in a particularly dramatic way. It also suggests that a requirement to discover today something new in this area might well be the modesty needed to admit the equal contributions of both concerned communities.

After the pioneering time of the botanist Brown and the mathematician Bachelier, the concept of Brownian motion became familiar to the physicists thanks to the works of Einstein, Smoluchowski, Langevin, Ornstein and Uhlenbeck among others. Then Lévy, Wiener, Doob and Itô gave mathematical substance to these somewhat formal physical theories. Notice that all of them were motivated by the need of classical physics. After the quantum revolution, but with at least a major exception, to be specified later, physicists lost interest in the subject, till 1949 when Feynman published his work on the path integral formulation of non-relativistic quantum mechanics. His (formal) result was that it is possible to avoid the conventional description of the quantum dynamics in terms of the evolution group, i.e. of the one-parameter group of unitary operators on the Hilbert space $H = L^2(\mathbb{R})$ (with scalar product $\langle\cdot|\cdot\rangle$),

$$U_r = e^{-irH_\hbar/\hbar} \quad , \quad r \in \mathbb{R} , \tag{1.1}$$

where the Hamiltonian $H_\hbar$ is the self adjoint operator on a dense domain

$$H_\hbar = -\frac{\hbar^2}{2}\Delta + V , \tag{1.2}$$

V the multiplication by the (real)potential $V(x)$ and $\hbar$ the Planck constant. Instead of this, Feynman expresses the solution

$$(U_{r+\frac{s}{2}}\psi_{-s/2})(y)$$

of the Schrödinger equation (the time derivative of $\psi_r = U_{r+\frac{s}{2}}\psi_{-s/2}$) by a "Sum over Histories" in such a way that [1]

$$<\psi_{-s/2}|e^{-i(r+s/2)H_\hbar/\hbar}\psi_{-s/2}> = \iint \psi_{-s/2}(x)K(x,-s/2,y,r)\overline{\psi_{-s/2}(y)}dxdy$$

$$= (\mathrm{cste})\int_\Omega \psi_{-s/2}(\omega(-s/2))e^{i\int_{-s/2}^{r} L(\dot\omega(s'),\omega(s'))ds'}\overline{\psi_{-s/2}(\omega(r))}\mathcal{D}\omega \qquad (1.3)$$

where the (weak) integral kernel of $U(r+s/2)$, denoted by K, is called the propagator.

Ω is the space of all the paths $\Omega = \{\omega:[-s/2,r] \to \mathbb{R}\}$, $\mathcal{D}\omega = \prod_{-s/2\le s'\le r} dx(s')$ is regarded as a measure on Ω and L is the Lagrangian of the corresponding classical system (here a one-dimensional particle of unit mass in the force field $-\nabla V$),

$$L(\dot\omega,\omega) = \tfrac{1}{2}|\dot\omega|^2 - V(\omega) \; . \qquad (1.4)$$

Feynman's approach had a tremendous impact on theoretical physics and was at the origin of innumerable mathematical works. Among the "difficulties" to get over in order to give a sense to (1.3) there is the fact that the action functional $S[\omega(\cdot)] \equiv \int_{-s/2}^{r} L(\dot\omega,\omega)(s')ds'$ is only defined for regular paths $s' \to \omega(s')$ whereas, as shown by Feynman himself, the relevant quantum mechanical "paths" have the irregularities of the brownian trajectories. Moreover $\mathcal{D}\omega$ is not a measure. It has been shown, since then, that it is indeed possible to come pretty close, in a rigorous way, to (1.3) [2]. But each time, something is lost to develop further and successfully the analogy with quantum dynamics. In 1950, M. Kac discovered that the analytical continuation in time $r \to -it$ for U_r, which is possible under some restrictions on V, transforms the original problem is a solvable one, in the sense that the following path integral representation holds [3]:

$$<\psi_{-T/2}|e^{-(t+T/2)H_\hbar/\hbar}\psi_{-T/2}> = \iint \psi_{-T/2}(x)h(x,-T/2,y,t)\overline{\psi_{-T/2}(y)}dxdy$$

$$= \int_\Omega \psi_{-T/2}(\omega(-T/2))e^{-\int_{-T/2}^{t} V(\omega(s))ds}\overline{\psi_{-T/2}(\omega(t))}\, d\mu(\omega) \; . \qquad (1.5)$$

The key difference with (1.3) is that μ is a positive measure, namely the Wiener measure [4] with Lebesgue measure as initial condition. The integral kernel of $T_t = e^{-tH_\hbar/\hbar}$, denoted by h, is positive.

T_t is a contraction semigroup on H and the connection with quantum mechanics is built on the fact that, as an element of $L^2(\mathbb{R})$, the solution of the heat equation (the time derivative of $\theta^*_t = T_{t+T/2}\,\theta^*_{-T/2}$) is analytic in $\mathrm{Re}\, t > 0$, continuous for $\mathrm{Re}\, t \ge 0$ and its value on the imaginary axis solves the Schrödinger equation.

(1.5) is still today the (nonrelativistic) basis for Euclidean field theory. The difficulties of (1.5) are conceptual: the heat equation describes, in principle, only irreversible phenomena, qualitatively completely different from quantum ones, and the associated probabilistic interpretation of the dynamics (V as a "killing term") is physically meaningless. Also notice that, although the kinetic energy $\frac{1}{2}|\dot{\omega}|^2$ is absorbed in the measure of (1.5), its presence is as paradoxical as in (1.3). Physically speaking the Feynman-Kac formula (1.5) is a formula looking for a theory. E. Nelson proposed in 1966 (after Fényes in 1952) another, apparently unrelated, method to associate diffusion processes without killing with solutions of the Schrödinger equations [5].

The "major exception" mentioned before refers to E. Schrödinger himself. In 1931, unsatisfied by the current interpretation of quantum mechanics, he proposed the following subject of investigation [6]:

A system of classical particles of unit mass diffuses in a medium of diffusion constant λ, under the effect of a force field $F = -\nabla V$. It is assumed that these phenomena are described by the classical heat equation

$$- \lambda \frac{\partial \theta^*}{\partial t} = - \frac{\lambda^2}{2} \Delta\theta^* + V\theta^* \equiv H_\lambda \theta^* \tag{1.6}$$

namely the same as before, except for the diffusion coefficient λ and the fact that $\theta^* : \mathbb{R} \times I \to \mathbb{R}$, with t in $I = [-T/2,T/2]$, and some reasonable regularity conditions. The usual point of view is to regard (1.6) as a Cauchy problem. Schrödinger proposes to construct and to analyze the dynamical theory (if any) of the (classical) probabilistic evolutions associated with (1.6) and compatible with the data of <u>two</u> (arbitrary) <u>probability</u> densities,

$$p_{-T/2}(dx) = p_{-T/2}(x)dx \quad \text{and} \quad p_{T/2}(dy) = p_{T/2}(y)dy . \tag{1.7}$$

This program was sketched by Schrödinger in his original paper, then completely neglected by theoretical physicists. Few distinguished mathematicians were able to show its kinematical consistency [7].

The aim of this article is to summarize the dynamical solution of Schrödinger's problem [8]. It mainly results in a new probabilistic interpretation of the heat equation (1.6), more relevant than (1.5) as a classical analogue of Feynman's Sum over Histories.

2. Bernstein processes; outline of the probabilistic construction

Let M be a locally compact metric space, $\dot{M}$ its one point compactification. In the product topology $\Omega = \prod_{t\in I} \dot{M}$, $I = [-T/2,T/2]$, the path space is compact. A process is defined by $Z_t : \Omega \to \dot{M}$ $\omega \to \omega(t) \equiv Z_t(\omega)$. The Borel sigma-algebra of Ω

is $\sigma_I = \sigma\{Z_t, t \in I\}$.

Z_t on (Ω, σ_I, P) is a Bernstein process (1932, [7]) iff

$$E[f(Z_t) \mid \mathcal{P}_s \vee \mathcal{F}_u] = E[f(Z_t) \mid Z_s, Z_u] \tag{2.1}$$

for any $-T/2 < s < t < u < T/2$, any bounded Borel measurable function f . Here $\mathcal{P}_s$ denotes the past at time s , $\mathcal{P}_s = \sigma\{Z_s, s \le t\}$, $\mathcal{F}_u$ is the future at time u , $\mathcal{F}_u = \sigma\{Z_v , v \ge u\}$.

In modern terms, (2.1) is a one-dimensional version of the local Markov property. The class of Bernstein processes is strictly larger than the class of Markovian ones. Bernstein has to be credited for this early discovery.

Let us denote by B the Borel sigma algebra of M . By definition, a Bernstein transition H satisfies

(B1) $\forall\, x,y$ in M , $-T/2 \le s < t < u \le T/2$ the set function $A \to H(s,x;t,A;u,y)$ is a probability measure on B .

(B2) For any fixed set A in B and $-T/2 \le s < t < u \le T/2$, $(x,y) \to H(s,x;t,A;u,y)$ is $B \times B$ measurable.

(B3) For any A_1, A_2 in B , $-T/2 \le s < t < u < v \le T/2$,

$$\int_{A_2} H(s,w;t,A;u,y)H(s,w;u,dy,v,z) = \int_{A_1} H(s,w;t,dx;v,z)H(t,x;u,A_2;v,z) .$$

Theorem 1

Let M be as before, $H = H(s,x,t,A;u,y)$ a Bernstein transition, m a probability measure on $B \times B$. Then there is an unique probability measure P_m such that, with respect to (Ω, σ_I, P_m) , Z_t , $t \in I$ is a Bernstein process and

a) $P_m(Z_{-T/2} \in B_s , Z_{T/2} \in B_E) = m(B_s \times B_E)$, B_s and B_E in B

b) $P_m(Z_t \in B \mid Z_s, Z_u) = H(s,Z_s;t,B;u,Z_u) \quad \forall\ -T/2 < s < t < u < T/2$

c) The finite dimensional distribution of Z_t is given by

$$P_m(Z_{-T/2} \in B_s, Z_{t_1} \in B_1 ,\dots, Z_{t_n} \in B_n, Z_{T/2} \in B_E) =$$

$$\int_{B_s \times B_E} dm(x,y) \int_{B_1} H(-T/2,x,t_1,dx_1,T/2,y) \times \dots \int_{B_n} H(t_{n-1},x_{n-1};t_n,dx_n;T/2,y) .$$

Sketch of proof: (Jamison, [7])

For the uniqueness, suppose the existence of another measure P for Z_t , such that a) and b) hold. Then one shows by induction on n , and using the Bernstein property, that

$$P(Z_{t_2} \in B_2, \ldots, Z_{t_n} \in B_n \mid Z_{t_1}, Z_{T/2}) =$$

$$= \int_{B_2} H(t_1, Z_{t_1}; t_2, dz_2, T/2, Z_{T/2}) \int_{B_3} H(t_2, Z_2; t_3, dz_3; T/2, Z_{T/2})$$

$$\ldots \int_{B_n} H(t_{n-1}, Z_{n-1}; t_n, dz_n; T/2, Z_{T/2}) \quad .$$

The use of this in

$$P(Z_{-T/2} \in B_s, Z_{t_1} \in B_1, \ldots, Z_{t_n} \in B_n, Z_{T/2} \in B_E) = E[\chi_{Bs}(Z_{-T/2}) P(Z_{t_1} \in B_1, \ldots,$$

$$\ldots Z_{t_n} \in B_n \mid Z_{-T/2}) \chi_{B_E}(Z_{T/2})]$$

yields c) and therefore the uniqueness of P_m .

The existence proof of P_m has two equivalent versions. In the first one, the final position $Z_{T/2}$ is fixed during the construction, and in the second one, chosen here as an illustration, this is the initial position $Z_{-T/2}$. Consider $\Omega_{]-]} \equiv \{\omega :]-T/2, T/2] \to \dot{M}\}$, $\sigma_{]-]} = \sigma\{Z_t, t \in]-T/2, T/2]$ and, on the probability space of the boundary random variables, $\{\dot{M} \times \dot{M}, B \times B, m\}$, the two random variables $X(x,y) = x$ and $Y(x,y) = y$. By construction, for each x fixed in $\dot{M}$, $Q_{x*}(t,B,u,z) \equiv H(-T/2,x;t,B;u,z)$ is a (backward) Markovian transition probability. Defining $P_{T/2}(B_E, x) \equiv m(Y \in B_E \mid X = x)$, i.e. the disintegration of m , as final distribution, the theory of Markov processes yields the existence of the measure P_{x*} such that with respect to $(\Omega_{]-]}, \sigma_{]-]}, P_{x*})$, Z_t is a Markov process, denoted by Z_x , with this transition and final distribution. Now the distribution of the Bernstein

$Z_{-T/2}$ is required to be $\int_M \int_{B_s} m(x,y)dxdy$.

For the cylinder event $C_{]-]} = \{Z_{t_1} \in B_1, \ldots, Z_{t_n} \in B_n, Z_{T/2} \in B_E\}$,

$$P_m(B_s \times C_{]-]}) \equiv \int_{B_s} P_{x*}(C_{]-]}) \int_M m(x,y)dydx$$

is indeed the unique measure of Theorem 1. The rest of the proof is straightforward. □

In order to associate a Bernstein process to the heat equation (1.6), let us consider its integral kernel, $h(x,s,y,t+s) \equiv h(x,t,y) = \text{kernel } \{e^{-tH/\hbar}\}$. (2.2)

We restrict the class of potentials V in such a way that h is strictly positive. (For example, any continuous bounded below potential is in this class.)

Then for x,ξ,z in M , $-T/2 < s < t < u < T/2$,

$$h(s,x,t,\xi,u,y) = \frac{h(s,x,t,\xi)h(t,\xi,u,y)}{h(s,x,u,y)} \tag{2.3}$$

is the density of a Bernstein transition, i.e.

$$H(s,x,t,A,u,y) = \int_A h(s,x,t,\xi,u,y)d\xi \quad . \tag{2.3'}$$

In Schrödinger's problem, the data are a pair of boundary probabilities $p_{-T/2}(dx)$ and $p_{T/2}(dy)$. Many joint probabilities m , in the sense of Theorem 1, are compatible with these data. But only one choice of m yields a Bernstein process which is, moreover, Markovian:

<u>Theorem 2</u>

Let V be such that the integral kernel $h(x,t,y)$ is strictly positive and $B \times B$ measurable in (x,y) . Let $H(s,x,t,A,u,y)$ be the Bernstein transition defined by (2.3'), m a probability measure on $B \times B$ and Z_t , $t \in I$, the Bernstein process of the Theorem 1. Then

Z_t , $t \in I$, is also Markovian <=> $m \equiv M$ (M for Markovian) is of the form

$$M(B_S \times B_E) = \int_{B_S \times B_E} \overset{*}{\theta}_{-T/2}(x)h(x,T,y)\theta_{T/2}(y)dxdy \tag{2.4}$$

where $\overset{*}{\theta}_{-T/2}$ and $\theta_{T/2}$ are two bounded positive functions on M .

<u>Sketch of proof</u>

We summarize mainly <= , which is the constructive part of the proof.

After substitution of $m = M$ in the finite-dimensional distribution of Z_t ,

$$P_M(dx_1,t_1,\ldots,dx_n,t_n) =$$
$$= \int_{M\times M} \overset{*}{\theta}_{-T/2}(x)h(x,t_1+T/2,dx_1)\ldots h(dx_n,T/2-t_n,y)\theta_{T/2}(y)dxdy \tag{2.5}$$

where the notation $f(dx) = f(x)dx$ is used. Now define the forward evolution of $\overset{*}{\theta}_{-T/2}$ by

$$\overset{*}{\theta}(y,t) = \int_M \overset{*}{\theta}_{-T/2}(x)h(x,t+T/2,y)dx \qquad t \in I \tag{2.6}$$

(equivalently, $\overset{*}{\theta}_t(y) = \overset{*}{\theta}(y,t)$ is the positive solution of the Cauchy problem (2.6) with initial condition $\overset{*}{\theta}(y,-T/2) = \overset{*}{\theta}_{-T/2}(y)$).

Now define, for $A \in B$ and $-T/2 \le s \le t \le T/2$

$$Q_*(s,A,t,y) = \frac{1}{\theta^*(y,t)} \int_A \theta^*(x,s)h(x,t-s,y)dx \quad . \tag{2.7}$$

It is easy to show that Q_* is the backward transition probability of a Markov process and that (2.5) may also be written as

$$P_M(dx_1,t_1, \ldots, dx_n,t_n) =$$

$$= \int_M Q_*(t_1,dx_1,t_2,x_2)Q_*(t_2,dx_2,t_3,x_3) \ldots Q_*(t_n,dx_n,T/2,y)p(y,T/2)dy \tag{2.5'}$$

for the final distribution probability

$$p(dy,T/2) = \theta^*(y,T/2)\theta_{T/2}(y)dy \quad . \tag{2.8}$$

This (backward) Markovian process will be denoted by $Z^y(t)$ afterwards. A similar proof holds in terms of the backward evolution of $\theta_{T/2}$, defined by

$$\theta(x,t) = \int_M h(x,T/2-t,y)\theta_{T/2}(y)dy \quad , \qquad t \in I \tag{2.9}$$

if we identify Z_t with a (forward) Markovian process, denoted by $Z_x(t)$.

Equation (2.7) is reminiscent of Doob's construction of an h-path process [9]. Let us emphasize, however, that our starting kernel h is not, in general, the density of a conservative Markovian transition probability.

For => , one starts from the duality relation between the forward and backward transition probabilities of the Markovian process Z_t relative to its density and one compares the expressions of

$$P_m(Z_{-T/2} \in B_S, Z_t \in B, Z_{T/2} \in B_E)$$

where Z_t is first regarded as a Markov process and then as a Bernstein process with joint probability m .

Remark

A crucial consequence of this construction is that, for any $t \in I$, the probability density of Z_t is given by

$$p(x,t)dx = \theta^*(x,t)\theta(x,t)dx \ , \tag{2.10}$$

in striking analogy with a quantum probability density. Notice that $\theta_t = \theta(x,t), t \in I$, solves $\lambda\frac{\partial\theta}{\partial t} = H_\lambda\theta$ with $\theta(x,T/2) = \theta_{T/2}(x)$. In particular, this probability density is time symmetric. The relation (2.10) is at the origin of Schrödinger's idea [6].

In the particular case of a smooth Bernstein process on $M = \mathbb{R}$, it is immediate using (2.7), to find the local characteristic of $Z^y(t)$, namely the (backward) drift and diffusion coefficient,

$$B_{*}(y,t) = \lim_{\Delta t \downarrow 0} \frac{1}{\Delta t} \int_{S_{\varepsilon}(y)} (y-x) q_{*}(t-\Delta t,x,t,y)dx = - \frac{\nabla \theta^{*}}{\theta^{*}}(y,t) \tag{2.11}$$

and

$$C_{*}(y,t) = \lim_{\Delta t \downarrow 0} \frac{1}{\Delta t} \int_{S_{\varepsilon}(y)} (y-x)^{2} q^{*}(t-\Delta t,x,t,y)dx = \hbar \tag{2.12}$$

where $S_{\varepsilon}(y)$ is the sphere of center y and radius ε and $Q_{*}(s,dx,t,y) = q_{*}(s,x,t,y)dx$, for $-T/2 \le s \le t \le T/2$. The local characteristic of $Z_x(t)$, $t \in I$, follows along the same line. Notice that Theorem 2 neither specifies the positive functions $\theta^{*}_{-T/2}$ and $\theta_{T/2}$ nor the conditions for their existence. But the marginals of M yields the following system of nonlinear functional equations for these positive functions,

$$\begin{aligned} \theta^{*}_{-T/2}(x) \int_{M} h(x,T,y)\theta_{T/2}(y)dy &= p_{-T/2}(x) \\ \theta_{T/2}(y) \int_{M} \theta^{*}_{-T/2}(x) h(x,T,y)dx &= p_{T/2}(y) \end{aligned} \tag{2.13}$$

where $p_{-T/2}$ and $p_{T/2}$ are the data (1.7) of our problem. I shall call Equation (2.13) Schrödinger's system [6] :

Theorem 3

Let $p_{-T/2}(dx)$ and $p_{T/2}(dy)$ be strictly positive probability measures on B, for M a locally compact metric space. Let $h(x,T,y) \equiv h(x,y)$ be a continuous and strictly positive kernel on $M \times M$. Then the solutions of the Schrödinger system (2.13) exists and are unique.

Corollary

In these conditions, the measure of an unique Markovian Bernstein process is well defined.

The proof is an adaptation of a Theorem of Beurling [7]. One shows the existence of an unique pair (m,π) where m is a probability measure on $B \times B$ and π a sigma-finite product measure on $B \times B$ such that

$$\begin{aligned} dm(x,y) &= h(x,y)d\pi(x,y) \\ &= h(x,y)\, \theta^{*}_{-T/2}(x)\theta_{T/2}(y)dxdy \end{aligned}$$

and such that the conditions on the marginals on m are equivalent to (2.13).

3. Dynamical law

Given the local characteristics of the smooth Markovian Bernstein process of §2, simple calculations show that

$$D_* D_* Z^y(t) = \nabla V(Z^y(t)) \tag{3.1}$$

and, if D is defined by $DF(s) = \lim_{ds \downarrow 0} \frac{1}{ds} E_s[F(s+ds)-F(s)]$ for any P_s adapted $F(s)$,

$$DDZ_x(t) = \nabla V(Z_x(t)) \ . \tag{3.2}$$

Equations (3.1) and (3.2) will be interpreted as the dynamical laws of the theory, i.e. as the natural extensions of Newton equation in our context. But there are much better characterizations of these Bernstein processes.

In quantum mechanics, Feynman start from the (classical) Lagrangian (1.4). Since our construction involves the heat equation (1.6), it is natural to start from the Euclidean (classical) Lagrangian

$$\bar{L}(\omega,\dot{\omega}) \doteq \tfrac{1}{2}|\dot{\omega}|^2 + V(\omega) \ . \tag{3.3}$$

The presence of a formally infinite kinetic energy in both path integral representations of $U_r = e^{-irH_\hbar/\hbar}$ and $T_t = e^{-tH_\hbar/\hbar}$, Equations (1.3) and (1.5), suggests regularizing somewhat the classical concept of action functional. A diffusion process $Z^\varepsilon(s)$ is neighboring to a Bernstein $Z^0(s)$ if it is an $\mathbb{R}$-valued (for simplicity) Markovian F_s-semimartingale such that

$$\begin{aligned} d_* Z^\varepsilon(s) &= B_*^\varepsilon(Z\ (s),s)ds + d_* W_*(s) \qquad -T/2 \le s < t \le T/2 \\ Z^\varepsilon(t) &= y \end{aligned} \tag{3.4}$$

for $W_*(s)$ an F_s-martingale, and

$$B_*^\varepsilon(x,s) = -\left(\frac{\nabla\theta^*}{\theta^*}(x,s) + \varepsilon\nabla g^*(x,s)\right) \ . \tag{3.5}$$

In equation (3.5), θ^* is the solution of the Cauchy problem (1.6) with initial condition $\theta^*_{-T/2}$, g^* a smooth but arbitrary function on $\mathbb{R} \times I$ and ε a small real parameter. For $\varepsilon = 0$, $Z^\varepsilon(s) \equiv Z(s)$ coincides with the Markovian Bernstein process $Z^y(s)$ of §2. Notice that in Equation (3.4), d_* denotes the backward differential defined by $d_* F(s) = F(s)-F(s-ds)$ for any $F(s)$. To define a (backward) variation $\delta_* Z\ (s)$ we consider the derivative of the (backward) stochastic differential equation

$$Z^\varepsilon(s) = y + \int_t^s B_*^\varepsilon(Z^\varepsilon(r),r)dr + \int_t^s d_* W_*(r)$$

with respect to the parameter ε , at $\varepsilon = 0$. Then

$$\delta_\star Z(s) \equiv -\int_t^s \nabla g^\star(Z(r),r)dr + \int_t^s \nabla B_\star(Z(r),r)\delta_\star Z(r)dr \quad . \tag{3.6}$$

In particular,

$$D_\star\delta_\star Z(s) = -\nabla g^\star(Z(s),s) + \nabla B_\star(Z(s),s)\delta_\star Z(s) \quad . \tag{3.7}$$

The left hand side is actually a strong derivative since the process $\delta_\star Z$ is of bounded variation. Therefore this process is uniquely specified by the final boundary condition $\delta_\star Z(t) = 0$, and

$$Z^\varepsilon(s) = Z(s) + \varepsilon\delta_\star Z(s) + O(\varepsilon) \qquad -T/2 \le s \le t \le T/2 \tag{3.8}$$

with respect to the norm $||Z^\varepsilon - Z|| = \sup_{-T/2\le s\le t} E[|Z^\varepsilon(s)-Z(s)|^2]$. The action function is defined for any Z^ε by

$$A^*(Z^\varepsilon(s),s) = -\log \theta^*(Z^\varepsilon(s),s) \tag{3.9}$$

and solves

$$d_\star A^*(Z^\varepsilon(s),s) = D_\star A^*(Z^\varepsilon(s),s)ds + \nabla A^*(Z^\varepsilon(s),s)d_\star W_\star(s) \tag{3.10}$$

for $D_\star A^*(Z^\varepsilon(s),s) = \lim_{ds\downarrow 0} E_s \dfrac{A^*(Z^\varepsilon(s),s) - A^*(Z^\varepsilon(s-ds),s-ds)}{ds}$.

It has been shown in the first reference of [8] that

$$D_\star A^*(Z\ (s),s) \le \tfrac{1}{2}|B_\star^\varepsilon|^2(Z^\varepsilon(s),s) + V(Z^\varepsilon(s)) \tag{3.11}$$

and that the equality holds only for $\varepsilon = 0$, i.e. for $Z^\varepsilon(s) = Z^y(s)$ with $B_\star(Z^y(s),s) = \nabla A^*(Z^y(s),s)$. (Actually, (3.11) holds for a much larger class of admissible processes, including non markovian onces.) Equivalently, when the integral is finite,

$$A^*(Z^\varepsilon(t),t) - E_t A^*(Z^\varepsilon(-T/2),-T/2) \le E_t\int_{-T/2}^t \tfrac{1}{2}|B_\star^\varepsilon|^2(Z^\varepsilon(z),s) + V(Z^\varepsilon(s))\}ds \quad . \tag{3.11'}$$

Moreover, the comparison with the equality for $\varepsilon = 0$ yields

$$\delta A^*(Z(t),t) - E_t\delta A^*(Z(-T/2),-T/2) = E_t\{\int_{-T/2}^t -B_\star\nabla g^\star(Z(s),s)ds \; +$$
$$+ \int_{-T/2}^t (B_\star\nabla B_\star + \nabla V)(Z(s),s)\delta_\star Z(s)ds \quad .$$

But according to (3.7) this reduces to

$$\delta A^*(Z(t),t) - E_t\delta A^*(Z(-T/2),-T/2) = E_t\int_{-T/2}^t \{B_\star D_\star\delta_\star Z(s) + \nabla V(Z(s))\delta_\star Z(s)\}ds \tag{3.12}$$

and therefore, using (3.12), the inequality (3.11') and the fact that $A^*(Z^\varepsilon(s),s) = A^*(Z(s),s) + \varepsilon\delta A^*(Z(s),s) + 0(\varepsilon)$, the inequality (3.11') reduces to

$$E_t\int_{-T/2}^{t} \{\tfrac{1}{2}|B_*^\varepsilon|^2(Z^\varepsilon(s),s) + V(Z^\varepsilon(s))\}ds \geq E_t\int_{-T/2}^{t} \{\tfrac{1}{2}|B_*|^2(Z(s),s) + V(Z(s))\}ds$$
$$+ \varepsilon\, E_t\int_{-T/2}^{t} \{B_* D_* \delta_* Z(s) + \nabla V(Z(s))\delta_* Z(s)\}ds + 0(\varepsilon)\ . \qquad (3.13)$$

Let us define a stochastic action functional with initial condition (when it is finite) for an admissible Z^ε and $\bar{L}$ the Lagrangian (3.3) by

$$J[Z^\varepsilon(\cdot)] = E_t\int_{-T/2}^{t} \{\bar{L}(Z^\varepsilon(s),\ D_*Z^\varepsilon(s))\}ds \quad + \quad E_t A^*(Z^\varepsilon(-T/2),-T/2)\ . \qquad (3.14)$$

After an integration by parts in the r.h.s. of (3.13) and the use of $\delta_* Z(t) = 0$, we have

$$J[(Z^y + \varepsilon\delta_* Z^y)(\cdot)] - J[Z^y(\cdot)] \geq$$
$$\varepsilon\{E_t\int_{-T/2}^{t} \frac{\delta J}{\delta Z^y(s)}\ \delta_* Z(s)ds + E_t[(-B_* + \nabla A^*)\delta_* Z(-T/2)]\} + 0(\varepsilon) \qquad (3.13')$$

where the functional derivative $\dfrac{\delta J}{\delta Z^y(s)}$ is defined by

$$\frac{\delta J}{\delta Z^y(s)} = -\ D_* B_* + \nabla V\ . \qquad (3.14)$$

After division by ε, the r.h.s. of (3.13'), denoted by $\delta J[Z^y(\cdot)](\delta_* Z)$ is the (stochastic) Gâteau variation of J in direction $\delta_* Z$. Since (3.13') holds for any admissible variation $\delta_* Z(s)$, $-T/2 \leq s < t$, we have the

Theorem 4

Let $\bar{L}$ be the Euclidean Lagrangian (3.3). A smooth Markovian Bernstein process Z^y solution of the stochastic Newton equation

$$D_* D_* Z^y(s) = \nabla V(Z^y(s))\ , \qquad -T/2 \leq s < t \leq T/2 \qquad (3.15)$$

minimizes the action functional

$$J[Z(\cdot)] = E_t\int_{-T/2}^{t} \bar{L}(Z(s),D_* Z(s))ds + E_t A^*(Z(-T/2),-T/2) \qquad (3.16)$$

on the set of neighboring processes Z^ε such that $D_* Z^\varepsilon(-T/2) = \nabla A^*(Z^\varepsilon(-T/2),-T/2)$ and $Z^\varepsilon(t) = y$.

The given "mixed" boundary conditions (transversality conditions) are clearly sufficient for a minimal point.

Reciprocally, we can define Z as a local minimal for J if there is a positive

r such that $J[Z^{\varepsilon}] \geq J[Z]$ for all the Z^{ε} in the set of neighboring processes Y with $||Y-Z|| < r$. It follows that $\delta J[Z](\delta_* Z) = 0$, for any admissible variation $\delta_* Z$ such that $\delta_* Z(t) = 0$, is a necessary condition for a local minimum and this means precisely that Equation (3.15) holds with the boundary conditions $D_* Z(-T/2) = \nabla A^*(Z(-T/2),-T/2)$ and $Z(t) = y$. These conditions are therefore also necessary for the minimization.

Remarks

1) E. Nelson was the first to involve the drift $B_* = D_* Z$ in a kind of generalization of classical mechanics [5].

2) At the "classical limit" where the diffusion coefficient of the F_s-martingale $W_*(s)$ reduces to 0, Theorem 4 reduces to a well known Theorem of classical calculus of variation related to Hamilton-Jacobi's theory.

3) An analogous Theorem holds for the forward Markovian Bernstein process Z_x, involving the increasing filtration P_s and the forward drift $B = DZ$. The resulting Newton equation is Equation (3.2).

4) Theorem 4 describes the Markovian Bernstein process Z^y in terms of a conditional expectation, i.e. a transition probability. Such a description is essentially time assymetric. It is easy to find the dynamics of the Bernstein process Z_t, $t \in I$ along the lines of the two fixed endpoints problem of the classical calculus of variations. It is given by the (time symmetric) mean of the laws of motion for $Z^y(t)$ and $Z_x(t)$:

Theorem 5

The smooth Markovian Bernstein process of §2, $Z(t)$, $t \in I$, solves the stochastic Newton equation

$$\tfrac{1}{2}\,(D_* D_* Z + DDZ)(t) = \nabla V(Z(t)) \qquad t \in I \tag{3.17}$$

Remarks

1) Theorem 4 and 5 show that, for a natural stochastic extension of classical mechanics, the choice of boundary conditions has a qualitative influence over the dynamical description.

2) The law (3.17) contains all the dynamical information associated with the new probabilistic interpretations of the heat equation (1.6) summarized here. Notice that this one involves a particular class of solutions of (1.6) and yields time symmetric diffusion processes without killing (Bernstein processes).

3) It is possible to obtain directly (3.17) by Yasue's original stochastic variational principle [10] for the (finite) action functional

$$J[Z(\cdot)] = E[\int_{-T/2}^{T/2} \{\tfrac{1}{2}DZ.D_*Z + V(Z)\}dt] \quad . \tag{3.18}$$

4) In the frame proposed here, only functions of $Z(t)$ with symmetry under time reversal are natural. For example one defines $\bar{V} = \frac{1}{2}(B + B_*)$ and $\bar{U} = \frac{1}{2}(B - B_*)$ for B the (forward) drift of $Z_x(t)$. $\bar{V}$ is odd and $\bar{U}$ even under time reversal.

5) The initial gradient condition of Theorem 4 is preserved during the evolution on I so, using (2.11), $B_*(y,t) = \nabla A^*(y,t) = -\lambda \nabla \log \theta^*(y,t)$ and the definition (3.14) of the action functional we obtain the following path integral representation for our relevant solutions of the heat equation (1.6) :

Corollary

In the conditions of Theorem 4,

$$\theta^*(y,t) = \exp - \frac{1}{\lambda}\{E_t\int_{-T/2}^{t} \bar{L}(Z^y(s),D_*Z^y(s))ds + E_t A^*(Z(-T/2),-T/2)\} \quad . \tag{3.19}$$

4. Relations with quantum mechanics

Any Bernstein process, including the Markovian ones described in §2 and 3, is defined exclusively in terms of a pair of (strictly positive) probability densities, $p_{-T/2}(dx) = p_{-T/2}(x)dx$ and $p_{T/2}(dy) = p_{T/2}(y)dy$. Both are needed, and this is what gives to these diffusion processes their exceptional properties. But this also means that we cannot exhibit any explicit example of dynamical Bernstein process $Z(t)$, $t \in I$, without solving Schrödinger's system (2.13), and this is obviously rather hard.

Nevertheless, here is a simple way to produce a large class of such explicit dynamical Bernstein processes. We consider only the smooth case for simplicity, the regularity assumption will be specified in a forthcoming article (Third ref of [8]).

Suppose that we are given a (strict) solution ψ_r in $L^2(\mathbb{R})$, r in $[-s/2,s/2]$ of the Schrödinger equation

$$i\hbar \frac{\partial \psi}{\partial r} = H_\hbar \psi \tag{4.1}$$

for $H_\hbar$ defined in (1.2). We assume that the expectation of its energy is positive and finite, i.e.

$$0 < \int_{\mathbb{R}} \{\frac{\hbar^2}{2} |\nabla \psi_r|^2 + V|\psi_r|^2\}dx \equiv \langle \psi_r | H_\hbar \psi_r \rangle < \infty \quad . \tag{4.2}$$

Let us represent such a solution ψ_r by

$$\psi(x,r) = e^{(R+iS)(x,r)/\hbar} \tag{4.3}$$

or, more specifically, if v_0 denotes $\nabla S(x,0)$, by $\psi_{v_0}(x,r)$. Now let us introduce an analytical continuation in time of this solution, $t = ir$, $r > 0$, by

$$\begin{aligned}\theta^*(x,t) &= \psi_{-iv_0}(x,-it) \\ &\equiv e^{(\bar{R}-\bar{S})(x,t)/\hbar} \quad .\end{aligned} \tag{4.4}$$

Since the equations of evolution of S and R on $[-s/2,s/2]$ are known (by (4.1)), $\bar{R}$ and $\bar{S}$ solve the following real system of coupled non-linear partial differential equations on $\mathbb{R} \times I$,

$$\frac{\partial \bar{R}}{\partial t} = -\frac{\hbar}{2}\Delta\bar{S} - \nabla\bar{R}.\nabla\bar{S} \tag{4.5}$$

$$\frac{\partial \bar{S}}{\partial t} = -\frac{\hbar}{2}\Delta\bar{R} - \tfrac{1}{2}(\nabla\bar{R})^2 - \tfrac{1}{2}(\nabla\bar{S})^2 + V \quad . \tag{4.6}$$

Taking gradients and using the notations of the third remark after Theorem 5, for

$$\bar{V} \equiv \nabla\bar{S} \quad , \quad \bar{U} \equiv \nabla\bar{R} \; , \tag{4.7}$$

one shows easily that Equation (4.6) is nothing but the law of motion (3.17) and that Equation (4.5) expresses the local conservation of probability for a Markovian Bernstein process. Analogously,

$$\theta(x,t) = e^{(\bar{R}+\bar{S})(x,t)/\hbar} \quad . \tag{4.4'}$$

This means that to any regular solution of the Schrödinger equation is associated, after analytical continuation in time, an unique Markovian Bernstein process $Z(t)$ for the pair of probability densities

$$\begin{aligned}p_{-T/2}(x)dx &= e^{2\bar{R}(x,-T/2)}\,dx \\ p_{T/2}(y)dy &= e^{2\bar{R}(y,T/2)}\,dy\end{aligned} \tag{4.8}$$

when they fulfil the conditions (Theorem 3) of existence and uniqueness of the solutions for the associated Schrödinger system (2.13).

The Euclidean analog of the constraint (4.2) is

$$0 < e = \int_{\mathbb{R}} (\tfrac{1}{2}\,\bar{U}^2 - \tfrac{1}{2}\,\bar{V}^2 + V)(x,t)p(x,t)dx \tag{4.9}$$

where e is a time independent real number (a conserved quantity). The minus sign in this integral allows compensation of arbitrary large $\bar{U}$ and $\bar{V}$ (that is B and B_*) for fixed e. So, in general, we shall have explosion of the drifts of $Z(t)$ in a finite time. This new dynamical theory of Brownian motion is local in time.

If we come back into real time, by $\mathbb{R}$ $r = -it$, we get another dynamical theory of Brownian motion associated with the Schrödinger equation (4.1), and another

Markovian diffusion, $X(r)$, $r \in \mathbb{R}$. For this process, the analog of (4.9) involves only positive signs in the conserved energy (cf.(4.2)) and the theory is global in time. This is Nelson's stochastic mechanics [5] and the theory presented here, inspired by Schrödinger, is the imaginary time version of Nelson's theory whose existence was, at the very least, unexpected. Except in the particular case of stationary states, these two theories are very different.

5. Conclusion and prospects

The new probabilistic interpretation of the heat equation summarized here may present some interest for mathematicians, since it suggests a lot of fresh directions of investigation. But our main motivation, as the one of Schrödinger, comes from theoretical physics.

This theory allows us to reconsider the basic problem of the relations between quantum mechanics and probability theory. In the two last references of [8] it is shown that there is a new Hilbert space $\bar{H}$ of real functions on $\mathbb{R}$ (for ex.), with an internal operation denoted by $*$ and compatible with the above construction. In a natural way, any quantum mechanical operators on $L^2(\mathbb{R})$ can be transferred onto a (densely defined) linear operator on $\bar{H}$. If $(\cdot|\cdot)$ denotes the scalar product on $\bar{H}$, we have, in particular,

$$(\theta^*_{-T/2}|e^{-(t+T/2)H_\lambda/\lambda}\,\theta^*_{-T/2}) = \iint \theta^*_{-T/2}(x)h(x,t+T/2,y)\theta_{-T/2}(y)dxdy \quad . \tag{5.1}$$

The cone of the positive functions in $\bar{H}$ is the analogue of quantum states and is associated with the probabilistic interpretation summarized here.

The structure of this theory is incomparably closer of the structure of quantum mechanics than in the approaches inspired by Feynman-Kac formula (1.5). As depicted by Th. 4 and 5, at the classical limit of (strongly) differentiable trajectories $t \to Z^y(t)$, the theory reduces to (Euclidean) classical mechanics.

The prospect is that the solution of Schrödinger's problem is the genuine Euclidean version of quantum mechanics, and will be more relevant as a starting point for Euclidean field theory.

Acknowledgments

It is a pleasure to thank the organizing committee of this conference, and in particular Professor A. Truman, for their warm hospitality.

The ideas presented here have greatly benefited from friendly discussions with Professors S. Albeverio and E. Nelson and from collaboration with Professor K. Yasue. I thank particularly Professor D. Elworthy for a number of very useful comments regarding this paper. They result, for example, in a real improvement of Theorem 4.

Partially supported by SERC Grant N° GRD23404.

References

[1] R.P. Feynman, Rev. Mod. Phys., 20, 267 (1948); with A.R. Hibbs, "Quantum Mechanics and Path Integrals", McGraw-Hill N.Y. (1965).

[2] E. Nelson, J. Math. Phys. 5, 332 (1964); S. Albeverio and R. Hoegh-Krohn, "Mathematical Theory of Feynman Path Integrals", Lect. Notes in Math. 523, Springer-Verlag, Berlin (1976); D. Elworthy and A. Truman, Ann. Inst. Henri Poincaré Vol. 41, n° 2, 115 (1984).

[3] M. Kac, "On some connections between probability theory and differential and integral equations" in Proc. of 2st Berkeley Symposium on Probability and Statistics, J. Neyman Ed., Univ. of California Press, Berkeley (1951).

[4] N. Wiener, J. Math. Phys. 2, 132 (1923).

[5] I. Fenyes, Zeitsch. für Phys., 132, 81 (1952)
E. Nelson, Phys. Rev. 150, 1079 (1966); "Quantum Fluctuations", Princeton Univ. Press (1985).

[6] E. Schrödinger, Ann. Inst. Henri Poincaré 11, 300 (1932).

[7] S. Bernstein, "Sur les liaisons entre les grandeurs aléatoires", Verh des intern. Mathematikerkongr., Zurich, Band 1 (1932); R. Fortet, J. Math. Pures et Appl. IX, 83 (1940); A. Beurling, Annals of Math., 72, 1, 189 (1960); B. Jamison, Z. Wahrscheinlich, ver Gebiete 30, 65 (1964).

[8] J.C. Zambrini, J. Math. Phys. 27, 9, 2307 (1986); "Euclidean Quantum Mechanics, Phys. Rev. A, 35, 9, 3631 (1987); with S. Albeverio and K. Yasue "Hilbert space approach to Euclidean quantum mechanics", to appear.

[9] J.L. Doob, A Markov chain Theorem. Probability and Statistic, in The Harold Cramér Volume, N.Y. Wiley (1959).

[10] K. Yasue, J. Math. Phys. 22, 1010 (1981); J. Funct. Anal. 41, 327 (1981).

Vol. 1173: H. Delfs, M. Knebusch, Locally Semialgebraic Spaces. XVI, 329 pages. 1985.

Vol. 1174: Categories in Continuum Physics, Buffalo 1982. Seminar. Edited by F.W. Lawvere and S.H. Schanuel. V, 126 pages. 1986.

Vol. 1175: K. Mathiak, Valuations of Skew Fields and Projective Hjelmslev Spaces. VII, 116 pages. 1986.

Vol. 1176: R.R. Bruner, J.P. May, J.E. McClure, M. Steinberger, H_∞ Ring Spectra and their Applications. VII, 388 pages. 1986.

Vol. 1177: Representation Theory I. Finite Dimensional Algebras. Proceedings, 1984. Edited by V. Dlab, P. Gabriel and G. Michler. XV, 340 pages. 1986.

Vol. 1178: Representation Theory II. Groups and Orders. Proceedings, 1984. Edited by V. Dlab, P. Gabriel and G. Michler. XV, 370 pages. 1986.

Vol. 1179: Shi J.-Y. The Kazhdan-Lusztig Cells in Certain Affine Weyl Groups. X, 307 pages. 1986.

Vol. 1180: R. Carmona, H. Kesten, J.B. Walsh, École d'Été de Probabilités de Saint-Flour XIV – 1984. Édité par P.L. Hennequin. X, 438 pages. 1986.

Vol. 1181: Buildings and the Geometry of Diagrams, Como 1984. Seminar. Edited by L. Rosati. VII, 277 pages. 1986.

Vol. 1182: S. Shelah, Around Classification Theory of Models. VII, 279 pages. 1986.

Vol. 1183: Algebra, Algebraic Topology and their Interactions. Proceedings, 1983. Edited by J.-E. Roos. XI, 396 pages. 1986.

Vol. 1184: W. Arendt, A. Grabosch, G. Greiner, U. Groh, H.P. Lotz, U. Moustakas, R. Nagel, F. Neubrander, U. Schlotterbeck, One-parameter Semigroups of Positive Operators. Edited by R. Nagel. X, 460 pages. 1986.

Vol. 1185: Group Theory, Beijing 1984. Proceedings. Edited by Tuan H.F. V, 403 pages. 1986.

Vol. 1186: Lyapunov Exponents. Proceedings, 1984. Edited by L. Arnold and V. Wihstutz. VI, 374 pages. 1986.

Vol. 1187: Y. Diers, Categories of Boolean Sheaves of Simple Algebras. VI, 168 pages. 1986.

Vol. 1188: Fonctions de Plusieurs Variables Complexes V. Séminaire, 1979–85. Edité par François Norguet. VI, 306 pages. 1986.

Vol. 1189: J. Lukeš, J. Malý, L. Zajíček, Fine Topology Methods in Real Analysis and Potential Theory. X, 472 pages. 1986.

Vol. 1190: Optimization and Related Fields. Proceedings, 1984. Edited by R. Conti, E. De Giorgi and F. Giannessi. VIII, 419 pages. 1986.

Vol. 1191: A.R. Its, V.Yu. Novokshenov, The Isomonodromic Deformation Method in the Theory of Painlevé Equations. IV, 313 pages. 1986.

Vol. 1192: Equadiff 6. Proceedings, 1985. Edited by J. Vosmansky and M. Zlámal. XXIII, 404 pages. 1986.

Vol. 1193: Geometrical and Statistical Aspects of Probability in Banach Spaces. Proceedings, 1985. Edited by X. Femique, B. Heinkel, M.B. Marcus and P.A. Meyer. IV, 128 pages. 1986.

Vol. 1194: Complex Analysis and Algebraic Geometry. Proceedings, 1985. Edited by H. Grauert. VI, 235 pages. 1986.

Vol. 1195: J.M. Barbosa, A.G. Colares, Minimal Surfaces in $\mathbb{R}^3$. X, 124 pages. 1986.

Vol. 1196: E. Casas-Alvero, S. Xambó-Descamps, The Enumerative Theory of Conics after Halphen. IX, 130 pages. 1986.

Vol. 1197: Ring Theory. Proceedings, 1985. Edited by F.M.J. van Oystaeyen. V, 231 pages. 1986.

Vol. 1198: Séminaire d'Analyse, P. Lelong – P. Dolbeault – H. Skoda. Seminar 1983/84. X, 260 pages. 1986.

Vol. 1199: Analytic Theory of Continued Fractions II. Proceedings, 1985. Edited by W.J. Thron. VI, 299 pages. 1986.

Vol. 1200: V.D. Milman, G. Schechtman, Asymptotic Theory of Finite Dimensional Normed Spaces. With an Appendix by M. Gromov. VIII, 156 pages. 1986.

Vol. 1201: Curvature and Topology of Riemannian Manifolds. Proceedings, 1985. Edited by K. Shiohama, T. Sakai and T. Sunada. VII, 336 pages. 1986.

Vol. 1202: A. Dür, Möbius Functions, Incidence Algebras and Power Series Representations. XI, 134 pages. 1986.

Vol. 1203: Stochastic Processes and Their Applications. Proceedings, 1985. Edited by K. Itô and T. Hida. VI, 222 pages. 1986.

Vol. 1204: Séminaire de Probabilités XX, 1984/85. Proceedings. Edité par J. Azéma et M. Yor. V, 639 pages. 1986.

Vol. 1205: B.Z. Moroz, Analytic Arithmetic in Algebraic Number Fields. VII, 177 pages. 1986.

Vol. 1206: Probability and Analysis, Varenna (Como) 1985. Seminar. Edited by G. Letta and M. Pratelli. VIII, 280 pages. 1986.

Vol. 1207: P.H. Bérard, Spectral Geometry: Direct and Inverse Problems. With an Appendix by G. Besson. XIII, 272 pages. 1986.

Vol. 1208: S. Kaijser, J.W. Pelletier, Interpolation Functors and Duality. IV, 167 pages. 1986.

Vol. 1209: Differential Geometry, Peñíscola 1985. Proceedings. Edited by A.M. Naveira, A. Ferrández and F. Mascaró. VIII, 306 pages. 1986.

Vol. 1210: Probability Measures on Groups VIII. Proceedings, 1985. Edited by H. Heyer. X, 386 pages. 1986.

Vol. 1211: M.B. Sevryuk, Reversible Systems. V, 319 pages. 1986.

Vol. 1212: Stochastic Spatial Processes. Proceedings, 1984. Edited by P. Tautu. VIII, 311 pages. 1986.

Vol. 1213: L.G. Lewis, Jr., J.P. May, M. Steinberger, Equivariant Stable Homotopy Theory. IX, 538 pages. 1986.

Vol. 1214: Global Analysis – Studies and Applications II. Edited by Yu.G. Borisovich and Yu.E. Gliklikh. V, 275 pages. 1986.

Vol. 1215: Lectures in Probability and Statistics. Edited by G. del Pino and R. Rebolledo. V, 491 pages. 1986.

Vol. 1216: J. Kogan, Bifurcation of Extremals in Optimal Control. VIII, 106 pages. 1986.

Vol. 1217: Transformation Groups. Proceedings, 1985. Edited by S. Jackowski and K. Pawalowski. X, 396 pages. 1986.

Vol. 1218: Schrödinger Operators, Aarhus 1985. Seminar. Edited by E. Balslev. V, 222 pages. 1986.

Vol. 1219: R. Weissauer, Stabile Modulformen und Eisensteinreihen. III, 147 Seiten. 1986.

Vol. 1220: Séminaire d'Algèbre Paul Dubreil et Marie-Paule Malliavin. Proceedings, 1985. Edité par M.-P. Malliavin. IV, 200 pages. 1986.

Vol. 1221: Probability and Banach Spaces. Proceedings, 1985. Edited by J. Bastero and M. San Miguel. XI, 222 pages. 1986.

Vol. 1222: A. Katok, J.-M. Strelcyn, with the collaboration of F. Ledrappier and F. Przytycki, Invariant Manifolds, Entropy and Billiards; Smooth Maps with Singularities. VIII, 283 pages. 1986.

Vol. 1223: Differential Equations in Banach Spaces. Proceedings, 1985. Edited by A. Favini and E. Obrecht. VIII, 299 pages. 1986.

Vol. 1224: Nonlinear Diffusion Problems, Montecatini Terme 1985. Seminar. Edited by A. Fasano and M. Primicerio. VIII, 188 pages. 1986.

Vol. 1225: Inverse Problems, Montecatini Terme 1986. Seminar. Edited by G. Talenti. VIII, 204 pages. 1986.

Vol. 1226: A. Buium, Differential Function Fields and Moduli of Algebraic Varieties. IX, 146 pages. 1986.

Vol. 1227: H. Helson, The Spectral Theorem. VI, 104 pages. 1986.

Vol. 1228: Multigrid Methods II. Proceedings, 1985. Edited by W. Hackbusch and U. Trottenberg. VI, 336 pages. 1986.

Vol. 1229: O. Bratteli, Derivations, Dissipations and Group Actions on C*-algebras. IV, 277 pages. 1986.

Vol. 1230: Numerical Analysis. Proceedings, 1984. Edited by J.-P. Hennart. X, 234 pages. 1986.

Vol. 1231: E.-U. Gekeler, Drinfeld Modular Curves. XIV, 107 pages. 1986.

Vol. 1232: P.C. Schuur, Asymptotic Analysis of Soliton Problems. VIII, 180 pages. 1986.

Vol. 1233: Stability Problems for Stochastic Models. Proceedings, 1985. Edited by V.V. Kalashnikov, B. Penkov and V.M. Zolotarev. VI, 223 pages. 1986.

Vol. 1234: Combinatoire énumérative. Proceedings, 1985. Edité par G. Labelle et P. Leroux. XIV, 387 pages. 1986.

Vol. 1235: Séminaire de Théorie du Potentiel, Paris, No. 8. Directeurs: M. Brelot, G. Choquet et J. Deny. Rédacteurs: F. Hirsch et G. Mokobodzki. III, 209 pages. 1987.

Vol. 1236: Stochastic Partial Differential Equations and Applications. Proceedings, 1985. Edited by G. Da Prato and L. Tubaro. V, 257 pages. 1987.

Vol. 1237: Rational Approximation and its Applications in Mathematics and Physics. Proceedings, 1985. Edited by J. Gilewicz, M. Pindor and W. Siemaszko. XII, 350 pages. 1987.

Vol. 1238: M. Holz, K.-P. Podewski and K. Steffens, Injective Choice Functions. VI, 183 pages. 1987.

Vol. 1239: P. Vojta, Diophantine Approximations and Value Distribution Theory. X, 132 pages. 1987.

Vol. 1240: Number Theory, New York 1984–85. Seminar. Edited by D.V. Chudnovsky, G.V. Chudnovsky, H. Cohn and M.B. Nathanson. V, 324 pages. 1987.

Vol. 1241: L. Gårding, Singularities in Linear Wave Propagation. III, 125 pages. 1987.

Vol. 1242: Functional Analysis II, with Contributions by J. Hoffmann-Jørgensen et al. Edited by S. Kurepa, H. Kraljević and D. Butković. VII, 432 pages. 1987.

Vol. 1243: Non Commutative Harmonic Analysis and Lie Groups. Proceedings, 1985. Edited by J. Carmona, P. Delorme and M. Vergne. V, 309 pages. 1987.

Vol. 1244: W. Müller, Manifolds with Cusps of Rank One. XI, 158 pages. 1987.

Vol. 1245: S. Rallis, L-Functions and the Oscillator Representation. XVI, 239 pages. 1987.

Vol. 1246: Hodge Theory. Proceedings, 1985. Edited by E. Cattani, F. Guillén, A. Kaplan and F. Puerta. VII, 175 pages. 1987.

Vol. 1247: Séminaire de Probabilités XXI. Proceedings. Edité par J. Azéma, P.A. Meyer et M. Yor. IV, 579 pages. 1987.

Vol. 1248: Nonlinear Semigroups, Partial Differential Equations and Attractors. Proceedings, 1985. Edited by T.L. Gill and W.W. Zachary. IX, 185 pages. 1987.

Vol. 1249: I. van den Berg, Nonstandard Asymptotic Analysis. IX, 187 pages. 1987.

Vol. 1250: Stochastic Processes – Mathematics and Physics II. Proceedings 1985. Edited by S. Albeverio, Ph. Blanchard and L. Streit. VI, 359 pages. 1987.

Vol. 1251: Differential Geometric Methods in Mathematical Physics. Proceedings, 1985. Edited by P.L. García and A. Pérez-Rendón. VII, 300 pages. 1987.

Vol. 1252: T. Kaise, Représentations de Weil et GL_2 Algèbres de division et GL_n. VII, 203 pages. 1987.

Vol. 1253: J. Fischer, An Approach to the Selberg Trace Formula via the Selberg Zeta-Function. III, 184 pages. 1987.

Vol. 1254: S. Gelbart, I. Piatetski-Shapiro, S. Rallis. Explicit Constructions of Automorphic L-Functions. VI, 152 pages. 1987.

Vol. 1255: Differential Geometry and Differential Equations. Proceedings, 1985. Edited by C. Gu, M. Berger and R.L. Bryant. XII, 243 pages. 1987.

Vol. 1256: Pseudo-Differential Operators. Proceedings, 1986. Edited by H.O. Cordes, B. Gramsch and H. Widom. X, 479 pages. 1987.

Vol. 1257: X. Wang, On the C*-Algebras of Foliations in the Plane. V, 165 pages. 1987.

Vol. 1258: J. Weidmann, Spectral Theory of Ordinary Differential Operators. VI, 303 pages. 1987.

Vol. 1259: F. Cano Torres, Desingularization Strategies for Three-Dimensional Vector Fields. IX, 189 pages. 1987.

Vol. 1260: N.H. Pavel, Nonlinear Evolution Operators and Semigroups. VI, 285 pages. 1987.

Vol. 1261: H. Abels, Finite Presentability of S-Arithmetic Groups. Compact Presentability of Solvable Groups. VI, 178 pages. 1987.

Vol. 1262: E. Hlawka (Hrsg.), Zahlentheoretische Analysis II. Seminar, 1984–86. V, 158 Seiten. 1987.

Vol. 1263: V.L. Hansen (Ed.), Differential Geometry. Proceedings, 1985. XI, 288 pages. 1987.

Vol. 1264: Wu Wen-tsün, Rational Homotopy Type. VIII, 219 pages. 1987.

Vol. 1265: W. Van Assche, Asymptotics for Orthogonal Polynomials. VI, 201 pages. 1987.

Vol. 1266: F. Ghione, C. Peskine, E. Sernesi (Eds.), Space Curves. Proceedings, 1985. VI, 272 pages. 1987.

Vol. 1267: J. Lindenstrauss, V.D. Milman (Eds.), Geometrical Aspects of Functional Analysis. Seminar. VII, 212 pages. 1987.

Vol. 1268: S.G. Krantz (Ed.), Complex Analysis. Seminar, 1986. VII, 195 pages. 1987.

Vol. 1269: M. Shiota, Nash Manifolds. VI, 223 pages. 1987.

Vol. 1270: C. Carasso, P.-A. Raviart, D. Serre (Eds.), Nonlinear Hyperbolic Problems. Proceedings, 1986. XV, 341 pages. 1987.

Vol. 1271: A.M. Cohen, W.H. Hesselink, W.L.J. van der Kallen, J.R. Strooker (Eds.), Algebraic Groups Utrecht 1986. Proceedings. XII, 284 pages. 1987.

Vol. 1272: M.S. Livšic, L.L. Waksman, Commuting Nonselfadjoint Operators in Hilbert Space. III, 115 pages. 1987.

Vol. 1273: G.-M. Greuel, G. Trautmann (Eds.), Singularities, Representation of Algebras, and Vector Bundles. Proceedings, 1985. XIV, 383 pages. 1987.

Vol. 1274: N. C. Phillips, Equivariant K-Theory and Freeness of Group Actions on C*-Algebras. VIII, 371 pages. 1987.

Vol. 1275: C.A. Berenstein (Ed.), Complex Analysis I. Proceedings, 1985–86. XV, 331 pages. 1987.

Vol. 1276: C.A. Berenstein (Ed.), Complex Analysis II. Proceedings, 1985–86. IX, 320 pages. 1987.

Vol. 1277: C.A. Berenstein (Ed.), Complex Analysis III. Proceedings, 1985–86. X, 350 pages. 1987.

Vol. 1278: S.S. Koh (Ed.), Invariant Theory. Proceedings, 1985. V, 102 pages. 1987.

Vol. 1279: D. Ieşan, Saint-Venant's Problem. VIII, 162 Seiten. 1987.

Vol. 1280: E. Neher, Jordan Triple Systems by the Grid Approach. XII, 193 pages. 1987.

Vol. 1281: O.H. Kegel, F. Menegazzo, G. Zacher (Eds.), Group Theory. Proceedings, 1986. VII, 179 pages. 1987.

Vol. 1282: D.E. Handelman, Positive Polynomials, Convex Integral Polytopes, and a Random Walk Problem. XI, 136 pages. 1987.

Vol. 1283: S. Mardešić, J. Segal (Eds.), Geometric Topology and Shape Theory. Proceedings, 1986. V, 261 pages. 1987.

Vol. 1284: B.H. Matzat, Konstruktive Galoistheorie. X, 286 pages. 1987.

Vol. 1285: I.W. Knowles, Y. Saitō (Eds.), Differential Equations and Mathematical Physics. Proceedings, 1986. XVI, 499 pages. 1987.

Vol. 1286: H.R. Miller, D.C. Ravenel (Eds.), Algebraic Topology. Proceedings, 1986. VII, 341 pages. 1987.

Vol. 1287: E.B. Saff (Ed.), Approximation Theory, Tampa. Proceedings, 1985–1986. V, 228 pages. 1987.

Vol. 1288: Yu. L. Rodin, Generalized Analytic Functions on Riemann Surfaces. V, 128 pages, 1987.

Vol. 1289: Yu. I. Manin (Ed.), K-Theory, Arithmetic and Geometry. Seminar, 1984–1986. V, 399 pages. 1987.